ROUTLEDGE LIBRARY EDITIONS:
POLLUTION, CLIMATE AND CHANGE

T0199364

Volume 2

THE IMPACT OF
MARINE POLLUTION

THE IMPACT OF
MARINE POLLUTION

Edited by
DOUGLAS J. CUSINE AND JOHN P. GRANT

Routledge
Taylor & Francis Group

LONDON AND NEW YORK

First published in 1980 by Croom Helm Ltd.

This edition first published in 2020
by Routledge
2 Park Square, Milton Park, Abingdon, Oxon OX14 4RN

and by Routledge
52 Vanderbilt Avenue, New York, NY 10017

Routledge is an imprint of the Taylor & Francis Group, an informa business

© 1980 Douglas J. Cusine and John P. Grant

British Library Cataloguing in Publication Data
A catalogue record for this book is available from the British Library

ISBN: 978-0-367-34494-8 (Set)
ISBN: 978-0-429-34741-2 (Set) (ebk)
ISBN: 978-0-367-35878-5 (Volume 2) (hbk)
ISBN: 978-0-367-35908-9 (Volume 2) (pbk)
ISBN: 978-0-429-34254-7 (Volume 2) (ebk)

Publisher's Note
The publisher has gone to great lengths to ensure the quality of this reprint but points out that some imperfections in the original copies may be apparent.

Disclaimer
The publisher has made every effort to trace copyright holders and would welcome correspondence from those they have been unable to trace.

The Impact of Marine Pollution

**EDITED BY DOUGLAS J. CUSINE
AND JOHN P. GRANT**

CROOM HELM   LONDON
ALLANHELD, OSMUN   MONTCLAIR

©1980 Douglas J. Cusine and John P. Grant
Croom Helm Ltd, 2-10 St John's Road, London SW11

British Library Cataloguing in Publication Data

The impact of marine pollution.
   1. Marine pollution – Law and legislation
   I. Cusine, Douglas J II. Grant, John P
   341.7'62   K3590.4

   ISBN 0–85664–933–3

Published in the United States of America in 1980
by Allanheld, Osmun & Co. Publishers, Inc.
19 Brunswick Road, Montclair, NJ 07042

Library of Congress Cataloging in Publication Data
Main entry under title:

The Impact of marine pollution.

   Includes index.
   1. Marine pollution – Law and legislation.
2. Oil pollution of the seas – Law and legislation.
I. Cusine, Douglas J. II. Grant, John P.
K3590.4.I66   341.7'62   80–670
ISBN 0–916672–54–9

Reproduced from copy supplied
printed and bound in Great Britain
by Billing and Sons Limited
Guildford, London, Oxford, Worcester

CONTENTS

INTERNATIONAL INSTRUMENTS

(arranged chronologically and with the popular name italicised)

International Convention for the Prevention of Pollution of the Sea by Oil 1954 (the *1954 Convention*), 1958 U.K.T.S. No. 56; Cmnd. 595; *New Directions in the Law of the Sea*, Vol. II, 557

International Convention on the Limitation of Liability of Owners of Seagoing Ships 1957, 1968 U.K.T.S. No. 52; Cmnd. 3678

Convention on the Territorial Sea and Contiguous Zone 1958, 1965 U.K.T.S. No. 3; Cmnd. 2511; *New Directions in the Law of the Sea*, Vol. I, 1

Convention on the High Seas 1958, 1963 U.K.T.S. No. 5; Cmnd. 1929, *New Directions in the Law of the Sea*, Vol. I, 257

Convention on the Continental Shelf 1958, 1964 U.K.T.S. No. 39; Cmnd. 2422; *New Directions in the Law of the Sea*, Vol. I, 101

International Regulations for Preventing Collisions at Sea 1960, 1966 U.K.T.S. No. 23; Cmnd. 2956; *New Directions in the Law of the Sea*, Vol. II, 509

International Convention for the Safety of Life at Sea 1960 (*SOLAS 1960*), 1965 U.K.T.S. No. 65; Cmnd. 2812

Paris Convention on Third Party Liability in the Field of Nuclear Energy 1960, 1968 U.K.T.S. No. 69; Cmnd. 3755

The 1962 Amendments to the International Convention for the Prevention of Pollution of the Sea by Oil (the *1962 Amendments*), 1967 U.K.T.S. No. 59; Cmnd. 3354; *New Directions in the Law of the Sea*, Vol. II, 567

Brussels Convention on the Liability of Operators of Nuclear Ships 1962, Leg/Conf/C.2/SR 13

Convention Supplementary to the Paris Convention on Third Party Liability in the Field of Nuclear Energy 1963, 1975 U.K.T.S. No. 44; Cmnd. 5948

International Convention on Load Lines 1968, 1968 U.K.T.S. No. 58; Cmnd. 3708

The 1969 Amendments to the International Convention for the Prevention of Pollution of the Sea by Oil (the *1969 Amendments*), 1978 U.K.T.S. No. 21; Cmnd. 7094; *New Directions in the Law of the Sea*, Vol. II, 580

International Convention on Civil Liability for Oil Pollution Damage

1969 (*CLC*), 1975 U.K.T.S. No. 106; Cmnd. 6183; (1970) 9 I.L.M. 45; *New Directions in the Law of the Sea*, Vol. II, 602

International Convention relating to Intervention on the High Seas in Cases of Oil Pollution Casualties 1969 (*Intervention Convention*), 1975 U.K.T.S. No. 77; Cmnd. 6056; (1970) 9 I.L.M. 25; *New Directions in the Law of the Sea*, Vol. II, 592

Tanker Owners' Voluntary Agreement concerning Liability for Oil Pollution 1969 (*TOVALOP*), (1969) 8 I.L.M. 497; *New Directions in the Law of the Sea*, Vol. II, 641

Vienna Convention on the Law of Treaties 1969, (1969) 8 I.L.M. 679

The 1971 Amendments to the International Convention for the Prevention of Pollution of the Sea by Oil (the *1971 Amendments*), 1972 U.K.T.S. Misc. No. 36; Cmnd. 5071; *New Directions in the Law of the Sea*, Vol. II, 589

International Convention for the Establishment of an International Fund for Compensation for Oil Pollution Damage 1971 (*Fund Convention*), 1978 U.K.T.S. No. 95; Cmnd. 7383; (1972) 11 I.L.M. 284; *New Directions in the Law of the Sea*, Vol. II, 611

Convention Relating to Civil Liability in the Field of Maritime Carriage of Nuclear Material 1971, (1972) 11 I.L.M. 277; *New Directions in the Law of the Sea*, Vol. II, 664

Treaty on the Prohibition of the Emplacement of Nuclear Weapons and Other Weapons of Mass Destruction on the Seabed and Ocean Floor and in the Subsoil 1971, *New Directions in the Law of the Sea*, Vol. I, 288

Contract Regarding an Interim Supplement to Tanker Liability for Oil Pollution 1971 (*CRISTAL*), (1971) 10 I.L.M. 137; *New Directions in the Law of the Sea*, Vol. II, 646

Convention on the Prevention of Marine Pollution by Dumping from Ships and Aircraft 1972 (*Oslo Dumping Convention*), 1975 U.K.T.S. No. 106; Cmnd. 6183; (1972) 11 I.L.M. 262

Convention on the Prevention of Marine Pollution by Dumping of Wastes and Other Matter 1972 (*London Dumping Convention*), 1978 U.K.T.S. No. 43; Cmnd. 6486; (1972) 11 I.L.M. 1291

Convention on International Regulations for Preventing Collisions at Sea 1972, 1977 U.K.T.S. No. 77; Cmnd. 6962; (1973) 12 I.L.M. 734; *New Directions in the Law of the Sea*, Vol. IV, 245

International Convention for the Prevention of Pollution from Ships 1973 (*MARPOL*), 1974 U.K.T.S. Misc. No. 26; Cmnd. 5748; (1973) 12 I.L.M. 1319; *New Directions in the Law of the Sea*, Vol. IV, 345

Protocol to the International Convention relating to Intervention on the High Seas in Cases of Oil Pollution Casualties 1973, 1975 U.K.T.S. Misc. No. 12; Cmnd. 6038; (1974) 13 I.L.M. 165; *New Directions in the Law of the Sea*, Vol. IV, 451

Convention on the Prevention of Marine Pollution from Land-Based Sources 1974, 1978 U.K.T.S. No. 64; Cmnd. 7251; (1974) 13 I.L.M. 352

International Convention for the Safety of Life at Sea 1974 (*SOLAS 1974*), (1975) 14 I.L.M. 959

Convention on the Protection of the Marine Environment of the Baltic Sea Area 1974 (*Helsinki Convention*), (1974) 13 I.L.M. 546

Offshore Pollution Liability Agreement 1975 (*OPOL*), (1974) 13 I.L.M. 1409 and (1975) 14 I.L.M. 147

Merchant Shipping (Improvement of Standards) Recommendation 1976, (1976) 15 I.L.M. 1293

Barcelona Convention for the Protection of the Mediterranean Sea against Pollution 1976 (*Barcelona Convention*), (1976) 15 I.L.M. 290

I.L.O. Convention on Minimum Standards in Merchant Ships 1976, (1976) 15 I.L.M. 1288.

Informal Composite Negotiating Text 1977 (*ICNT*), U.N. Doc. A./Conf/63/WP/10; (1977) 16 I.L.M. 1108

Convention on Civil Liability for Oil Pollution Damage from Offshore Operations 1976, (1977) 16 I.L.M. 1450

Convention on the Prohibition of the Development, Production and Stockpiling of Bacteriological and Toxic Weapons of Mass Destruction 1972, 1976 U.K.T.S. No. 11; Cmnd. 6397; (1972) 11 I.L.M. 309

Protocol to the International Convention for the Prevention of Pollution from Ships 1978, 1978 U.K.T.S. Misc. No. 27; Cmnd. 7347; (1978) 17 I.L.M. 546

Protocol to the International Convention for Safety of Life at Sea 1978, (1978) 17 I.L.M. 579

International Convention on Standards of Training, Certification and Watchkeeping for Seafarers 1978, IMCO Doc. Stw/Conf/13, 5 July 1978

ABBREVIATIONS AND DEFINITIONS

ACOPS	Advisory Committee on Oil Pollution of the Sea
A.C.	Appeal Cases
A.J.I.L.	American Journal of International Law
All E.R.	All England Law Reports
Art.	Article
barrel	1 Barrel is equivalent to 34.97261 Imperial gallons
B.Y.I.L.	British Yearbook of International Law
CBT	Clean Ballast Tanks
CLC	International Convention on Civil Liability for Oil Pollution Damage
Cmnd.	Command Paper (UK)
COW	Crude Oil Washing
CUEP	Central Unit on Environmental Pollution (UK)
Cur.Leg.Prob.	Current Legal Problems
D.	Dunlop, Court of Session Reports (Scotland)
D.T.I.	Department of Trade and Industry (UK)
dwt	Deadweight. Deadweight is the number of tons (of 2240 lb) required to sink a vessel to her load line. The deadweight therefore includes cargo, bunkers and stores
F.2d.	Federal Reporter, 2nd Series (USA)
F.Supp.	Federal Supplement (USA)
H.C. Deb.	House of Commons Reports (UK)
H.L. Deb.	House of Lords Reports (UK)
HMSO	Her Majesty's Stationery Office
I.B.A.	International Bar Association
I.C.L.Q.	International and Comparative Law Quarterly
I.C.J. Rep.	International Court of Justice Reports
IGS	Inert Gas System
I.L.A.	International Law Association
I.L.C.	International Law Commission
I.L.M.	International Legal Materials
I.L.O.	International Labour Organisation
IMCO	Inter-Governmental Maritime Consultative Organisation
J.B.L.	Journal of Business Law

J.M.L.C.	Journal of Maritime Law and Commerce
L.M.C.L.Q.	Lloyd's Maritime and Commercial Law Quarterly
LOT	Load-on-Top
N.L.J.	New Law Journal
O.J.	Official Journal (European Community)
P.C.I.J.	Permanent Court of International Justice
Q.B.	Queen's Bench Reports (England)
R.I.A.A.	(UN) Reports of International Arbitral Awards
S.	Section
SBT	Segregated Ballast Tanks
Sch.	Schedule
S.C.	Session Cases (Scotland)
S.D.L.R.	San Diego Law Review
S.I.	Statutory Instruments (UK)
S.Rep.	Senate Reports (USA)
tonne	A metric unit of weight equivalent to 0.984207 UK tons (Long Tons) and 1.10231 US tons (Short Tons)
T.I.S.C.	Trade and Industry Sub-Committee, Select Committee of the House of Commons (UK)
ULCC	Ultra Large Crude Carrier
U.K.T.S.	United Kingdom Treaty Series
UN	United Nations
U.S.	United States Supreme Court Reports
U.S.C.	United States Code
U.Tor.L.J.	University of Toronto Law Journal
VLCC	Very Large Crude Carrier
Yale L.J.	Yale Law Journal
Zaö R.V.	Zeitschrift Für Auslandisches öffentliches Recht und Völkerrecht

The following abbreviations are used for the Merchant Shipping Acts:

the 1894 Act: Merchant Shipping Act 1894
the 1906 Act: Merchant Shipping Act 1906
the 1949 Act: Merchant Shipping (Safety Convention) Act 1949
the 1958 Act: Merchant Shipping (Liability of Shipowners & Others) Act 1958
the 1965 Act: Merchant Shipping Act 1965
the 1971 Act: Merchant Shipping (Oil Pollution) Act 1971
the 1974 Act: Merchant Shipping Act 1974
the 1979 Act: Merchant Shipping Act 1979

PREFACE

The genesis of this book was the belief on the part of the editors that books on marine pollution tended to be of two sorts, those giving a general overview of the relevant international and municipal legal provisions, and those concentrating on one aspect, or incident, of marine pollution, often embodying not just legal, but technical and practical, considerations. The former type of book may be of interest only to students and those wishing an outline of the relevant law; the latter type of book may be of interest only to those with an active involvement in the oil industry, the shipping industry, or in governments or international institutions. The editors believe that there is a need to produce under one cover a book that to some extent bridges that gap.

The original intention was for the editors of this volume to produce the entire text themselves. While they might have been able to identify the relevant law, they were conscious that they were unqualified to explain the technical and practical problems of marine pollution and their resolution.

The editors decided, therefore, to invite a number of individuals with expertise in various aspects of marine pollution to contribute chapters to the book. The editors themselves wrote the introduction and chapter (Chapter 1) setting out in broad terms the general legal framework. These are intended to set the scene for the later chapters, and an attempt has been made to refer from them to the more detailed discussion of certain issues later in the book.

One of the notorious problems of collections of essays is that they tend to be piecemeal and unco-ordinated. The editors intended to attempt to make the whole work into a coherent, systematic whole. It soon became clear that this was unnecessary, for the contributions were broadly compatible. Any further attempt by the editors to co-ordinate the contributions would, it was thought, reduce the value of each contribution.

From the outset it was recognised that it would be impossible to include discussion of every aspect of marine pollution. Accordingly, the editors have been selective. As the vast bulk of the international agreements on pollution, which have subsequently been incorporated into municipal law, emerge from the Inter-Governmental Maritime Consultative Organisation, the work of IMCO merited consideration (Chapter 2). Conscious that there is now no dearth of international legislation on pollution, the authors thought it essential to include a contribution on what appears the major area of concern, the enforcement of these international standards (Chapter 3). The role of the oil companies and others as transporters of oil was another obvious area for inclusion (Chapter 4). It was initially thought necessary to include discussion of a number of important incidents, for example the *Torrey Canyon*, *Christos Bitas* and *Amoco Cadiz* incidents, but considerations of space and the realisation that these incidents would be considered in any event by the various contributors led to the conclusion that only the *Ekofisk Bravo* blow-out should be included (Chapter 5), largely to demonstrate the legal regime established for off-shore installations by Norway. Equally obvious, at least to the editors, was the role of insurance (Chapter 6).

While the main thrust of the book is clearly on marine pollution by oil, it was thought wise to include contributions on dumping of wastes at sea, on pollution from land-based sources and on the many problems caused by nuclear ships, nuclear cargo and the dumping of nuclear wastes (Chapters 7, 8 and 9). The editors wished to include some discussion of the legal rules applied in different parts of the world, and Europe and the United States were selected for inclusion (Chapters 10 and 11). However, throughout the book an attempt has been made to deal with the issues arising on an international and comparative basis, for it is abundantly clear that the resolution of most pollution problems cannot be achieved by each State acting on its own.

As this is not intended as a textbook, no table of cases or statutes has been included, but the UK Merchant Shipping Acts are referred to by abbreviations which are listed at the end of the list of Abbreviations and Definitions. A list of international instruments has been compiled to guide the reader to the source of the many international agreements, conventions and declarations that are referred to in the text, and to eliminate duplication of references to sources in the footnotes. The footnotes for each chapter are gathered at the end of the chapter.

The editors extend their gratitude to all the contributors, to Croom Helm, whose enthusiasm at times surpassed that of the editors and whose encouragement and assistance were unstinting; and to Mrs Fiona Chaplain of the University of Aberdeen and Miss Isabel Ballantyne of the University of Glasgow, who rendered the editors' illegible scrawl into typescript. The law is stated as at

1 March 1979 though it has been possible to take account of some later developments.

D.J. Cusine, Aberdeen
J.P. Grant, Glasgow

INTRODUCTION

> The last twelve months will be remembered as the period
> which witnessed the *Amoco Cadiz*, *Eleni V*, *Christos Bitas*,
> *Litiopa*, *Andros Patria*, *Esso Bernicia* and many other incidents.
> I have suggested in Parliament that we tend to practise
> Government by catastrophe, because policies are formulated
> only when disasters occur (Lord Ritchie Calder, *ACOPS Annual
> Report 1979*, p. 2).

> In 1978 alone we had the cases of the *Amoco Cadiz*, *Eleni V*
> and *Christos Bitas*, and it is only a matter of time before
> similar instances occur yet again (B. Sage, 'Bird Flare' in *Birds*
> Royal Society for Protection of Birds Magazine (Spring 1979)).

These statements are indicative of attitudes that are not at all uncommon
in relation to questions of pollution. It is often stated that pollution is
inevitable, incurable, catastrophic and inadequately attended to by
governments and commercial concerns. All in all, the impression is
given that pollution poses intractable problems, but this alleged intract-
ability may too easily disguise the real issues that fall to be resolved.

Take, for example, the stranding of the *Amoco Cadiz*. The *Amoco
Cadiz*, a Very Large Crude Carrier (VLCC) of 228,513 tons deadweight
(119,000 gross registered tons), was owned by the Amoco Transport
Company, was registered in Liberia and had an Italian crew. The vessel
was on charter to an affiliate of Royal Dutch Shell for a voyage from
the Arabian Gulf to Rotterdam, via Lyme Bay in the UK where she
was to off-load most of her cargo. She was carrying 120,000 tons of
light Iranian crude, 100,000 tons of light Arabian crude and several
tons of her own fuel oil.

On 16 March 1978, her steering failed off the island of Ushant and
she was carried by the current towards the north-west coast of Brittany.
After a delayed and ultimately unsuccessful attempt to tow the vessel
to sea, she stranded on the rocks off the Breton village of Portsall and
nearly all of the oil which she was carrying entered the sea, polluting
over 100 miles of the French coastline from Paimpol to Brest.

What, then, are the interests affected by the stranding of the *Amoco
Cadiz*? At a State level, both the exporting countries and the importing
countries were affected; for example, those countries for whom the

crude oil, once refined, was intended; the United States, as the country where the cargo owner was resident; Liberia, as the flag State. France suffered direct physical and economic damage. Her beaches were polluted, thus seriously affecting the tourist industry which provides a substantial income; the oil damaged or destroyed marine and bird life and local fishing activities, particularly oyster and lobster farming. Furthermore, she had to bear the initial cost of the extensive clean-up operation, which involved 3,000 French troops and many civilians; and French vessels were deployed to spray the sea with detergents over a considerable period, given that the bad weather and the condition of the vessel hampered their efforts. She also provided an initial compensation fund of just over $1 million.

Immediately affected were the fishermen, hoteliers and residents of Brittany. Less immediately affected must be the consumers of petroleum products, for the loss of 230,000 tons of oil diminished the overall supply and must lead to higher prices, however negligible those increases might be on the price of a gallon of petrol. The crew of the vessel were put at risk and, while no lives were lost, some careers may have been irreparably damaged. The masters of the *Amoco Cadiz* and the tug *Pacific* were arrested and charged with polluting the seas and the French Government blamed the master of the *Amoco Cadiz* for the incident.

Two hundred and thirty thousand tons of oil were lost to the owning company, as was the vessel, a not infrequent occurrence in an incident resulting in extensive oil pollution. As the shipowner, Amoco must have incurred (and presumably will incur in the future) considerable expenses in the stranding, in the subsequent clean-up operation and in the payment of compensation to the victims of the pollution, given the emphasis in the international conventions and the corresponding municipal legislation on the 'polluter-pays' principle. Some of these costs will be covered by insurance or by voluntary schemes within the oil and shipping industries, but in the case of the *Amoco Cadiz*, the available insurance was only $50m and, although the voluntary schemes would add another $36m, that leaves an enormous deficit. There have been various, and widely differing, estimates of the total costs involved, but by September 1978 the French Government had raised an action in New York claiming a total of $1,350m ($300m for the Government, $300m for local authorities and $750m for fishermen, hoteliers, traders, etc.). It has subsequently been estimated that the total costs may amount to $1,700m. There is thus a considerable amount for Amoco to find, and every claim settled must inevitably be reflected in an increase in the insurance premia which they will pay in the future.

Incidents like the *Amoco Cadiz* stranding attract widespread media coverage, and are the subject of public concern. Put simply, it is in nobody's interest — States, individuals or oil companies — to allow oil to enter the sea. Oil is an expensive (and diminishing) resource. If it escapes into the sea, it can cause damage to the amenity, to fishing grounds and to a tourist industry. Once in the sea, its effects have to be reduced by the use of mechanical lifting devices or dispersants, and these clean-up procedures are costly in terms of labour and equipment. Those who suffer as a result of oil pollution rightly demand that they be compensated at a high level. While the concern to prevent oil pollution may be shared equally, the means of achieving clean seas will, in most instances, be the subject of some degree of dispute between States and individuals on the one hand and the oil companies on the other hand.

However, the sea does not become polluted wholly and solely through accidents involving oil tankers. Off-shore oil extraction can result in immense pollution incidents; the *Ixtoc I* blow-out off Mexico's Yucatan Peninsula spewed oil into the Gulf of Mexico at the rate of 30,000 barrels a day from early June until September 1979, creating the largest single incident of oil pollution in history, and making the *Ekofisk Bravo* blow-out (with a total escape of 140,000 barrels) appear no more than a trickle. Vessels are used to transport wastes to be dumped at sea. Some vessels (not many at present) are powered by nuclear energy, others carry nuclear fuels and wastes; and the dangers inherent in these operations are obvious. A high proportion of the pollution of the marine environment comes from land-based sources, mainly in the form of industrial and domestic wastes put into watercourses. In all these types of marine pollution there are States, individuals and commercial interests which are affected, invariably adversely. In these instances, however, the various interest groups may align themselves differently: it may be in a State's interest, as it perceives it, to encourage wastes to be dumped at sea rather than on land, and to permit the fouling of watercourses with industrial wastes rather than require companies to provide expensive monitoring and treatment facilities. In situations in which a State is less than diligent in its commitment to the protection of the environment, it often falls to individuals, in particular, environmental groups, to provide the stimulus to remedial action.

It would be impossible within the compass of one book to offer definitive answers to the many economic, social, environmental, political and legal issues that surround each instance of marine pollution; but it is possible to set out some of the major issues, and to identify how States (through concerted or unilateral action) and oil companies respond to dangers to the marine environment.

Any thorough and comprehensive programme to protect the marine environment would involve a number of items, most of which are discussed in subsequent chapters of this book.

*Ensure that vessels are properly constructed (Chapters 1 and 3), adequately crewed (Chapters 1 and 3) and navigated in accordance with good practice and with national and international regulations (Chapter 1).

*Given that accidents are to some extent inevitable, ensure that satisfactory arrangements have been made to clean up the damage and to provide compensation for clean-up costs and for those who suffer as a result of oil pollution (Chapters 1, 4 and 6).

*Eliminate *harmful* discharge of oil through normal tanker operations (Chapter 1).

*Ensure that the construction, crewing and operation of off-shore installations are such as to prevent *harmful* escapes of oil into the sea, and to provide compensation for clean-up costs and for those who suffer pollution damage (Chapters 1, 5 and 6).

*Ensure that no *harmful* wastes are dumped at sea (Chapter 7).

*Ensure that untreated and *harmful* matter is not permitted to pollute the sea from land-based sources (Chapters 8 and 10).

*Ensure that the strictest standards are imposed upon those who operate nuclear-powered ships, and who carry nuclear fuels and waste (and other dangerous cargoes) (Chapter 9).

*Ensure international co-operation in combating pollution (Chapters 1 and 2); and ensure that regard is had to the law or practice of other States (Chapters 5, 10 and 11).

While it behoves no one, be they States, companies or individuals, to be complacent about pollution of the marine environment, it serves no useful purpose to fall into the inertia that invariably accompanies pessimism. It is clear that some pollution of the marine environment is inevitable and that some pollution, either because of its scale or its type, may be extremely harmful. However, the problems associated with marine pollution are now well known, and the identification of problems is the first step—indeed the prerequisite—towards resolving them. The risks can be, and have been, minimised. Measures can be, and have been, taken to mitigate the effects of pollution incidents when they occur. Compensation can be, and has been, made available to those who minimise pollution damage when it occurs and who suffer as a result of pollution. Whether what has been done to date is sufficient is a matter of judgement. Maintaining clean seas, like democracy, may be an issue requiring eternal vigilance.

Part I:
OIL POLLUTION

1 THE LEGAL FRAMEWORK

Douglas J. Cusine and John P. Grant

1 INTRODUCTION

The increasing concern within States about the protection of the environment has been matched, if not overtaken, by international concern about the pollution of the seas, particularly by oil. The estimated amount of oil entering the oceans annually is some 3.3m metric tons, of which 1.5m metric tons comes from ships, 1.7m metric tons from on-shore activities (including a massive 1.3m metric tons of discarded lubricants) and 0.08m metric tons from off-shore exploration and exploitation activities. A more detailed analysis of these figures can be seen from Table 1.[1]

Table 1: Estimate of Oil Entering the Oceans

	Metric Tons per Annum
Vessels	
Accidental	257,000
Operational/deliberate	
Deballasting and tank washing	
Using Load-on-Top	105,000
Non-Load-on-Top	529,000
Tank washing before maintenance	360,000
Bilge pumping	23,000
Bulk/oil carriers	46,000
Other ships	180,000
Off-shore Operations	
Accidental	80,000
Operational/deliberate	insignificant
Other Sources	
Tanker terminal operations	70,000
Refinery effluents	300,000
Pipelines and handling spillage	40,000
Discarded lubricants	1,300,000
Total	3,290,000

Pollution, by its nature, is not readily subject to the normal jurisdictional rules. If the flag State will not act to punish discharges of oil
from vessels of its nationality, the competence of States directly affected
by the discharge is restricted to the outer limits of territorial waters.
Canada, which was convinced of the inadequacy of these jurisdictional
rules, enacted the Canadian Arctic Waters Pollution Prevention Act in
1970,[2] purporting to require all vessels in a belt of waters extending up
to 100 miles from the Canadian coastline to comply with Canadian
regulations governing navigation, safety and the dumping of wastes.
This statute, which met with protests from the United States and United
Kingdom Governments, appears to go beyond what is permissible by
existing international law. It is clear, therefore, that pollution of the
sea can best be dealt with through international agreement, setting
standards for implementation and enforcement by States.[3]

It is sometimes appropriate in considering international law and
municipal law operating within the same field to deal first with international law (both customary law and conventional law) and then to
consider the position in municipal law (both at common law and under
statute). This approach is not considered wholly appropriate in issues
such as marine pollution. Instead, it is proposed to look at customary
international law and then the position at common law in the UK in an
attempt to demonstrate the ability or otherwise of these sources/systems/
regimes to provide a solution to the types of legal problem that invariably
arise from an incident causing marine pollution. Thereafter, we shall
analyse the international agreements concluded to prevent and to
mitigate the consequences of any such incident; and we shall analyse
the statutes giving effect to those agreements in one country – the UK.
To look at the international agreements and the UK statutes separately
would be quite inappropriate, as they are interrelated: the international
agreements were by and large concluded through an international
institution (IMCO) whose headquarters are in London and to whose
work the UK has been a consistent and strong contributor; and the UK
legislation that will be considered has by and large been passed to
implement these agreements. In short, the international agreements
and the UK legislation, while emanating from different sources and
having different effects, are closely related, and only by considering
them together can any reasonable assessment be made of their terms (for
the wording frequently differs) and their relative effectiveness.

Adopting that approach, our first question is therefore: does international law provide any rules of law on pollution in the absence of
international agreements? Put another way: are there any customary

rules of law on pollution, rules deriving their force from the common practice of States?[4] To recast the question yet again in a more practical way: are there any rules of international law which could bind those States that are not parties to some or all of the international agreements?

In one famous international arbitration, the *Trail Smelter Arbitration*,[5] there appears the following statement: 'no State has the right to use or permit the use of its territory in such a manner as to cause injury or damage . . . in or to the territory of another or the properties or persons therein.' This case concerned fumes, including quantities of sulphur dioxide, emitted from a lead and zinc smelter in Trail, British Columbia, which caused damage in the State of Washington. It is open to doubt how wide the *ratio* of the *Trail Smelter Arbitration* can be extended. On a narrow construction, it might be applicable to nothing more than damage caused in one State by activities carried out in another State. On a broad construction, it might be applicable to any damage caused to a State or its nationals by a vessel subject to the jurisdiction of another State. It is, of course, only on this latter construction that the case is relevant to the question of pollution of the sea by oil.

Some authorities take the view that the broad construction of the *ratio* of the *Trail Smelter Arbitration* is to be preferred.[6] This view is supported by the decision of the International Court of Justice in the *Corfu Channel Case*, where the Court recognised 'every State's obligation not to allow knowingly its territory to be used for acts contrary to the rights of other States';[7] by the International Law Commission's statement in 1956 that 'States are bound to refrain from any acts which might adversely affect the use of the high seas by nationals of other States';[8] and by some of the provisions of the Geneva Convention on the High Seas of 1958, which resulted from the work of the I.L.C. and which are expressly stated as being 'generally declaratory of established principles of international law', requiring the four freedoms of the high seas (navigation, fishing, laying cables and pipelines and overflight) to be exercised subject to other rules of law and to the rights of other States in the high seas, and imposing on all States a general obligation to prevent pollution of the seas.[9] Further, it can be argued that a broad construction of the *ratio* in the *Trail Smelter Arbitration* is more in accord with the nature of international law, a system that is generally thought to establish, through custom, broad principles of general application, rather than detailed rules to be followed in every particular.

In support of the contention that there is a customary legal regime governing pollution of the seas, reference may be made also to the doctrine of abuse of rights. This doctrine is based on the premiss that a

State is in breach of international law if it exercises a right that it has in
such a way as to prejudice other States exercising rights they enjoy. It
is a well established principle that the high seas are free and open for the
use of all States; that it is, in effect, *res communis*. Yet, each and every
State can only exercise the freedoms it has in the high seas, in the words
of the Convention on the High Seas, 'with reasonable regard to the
interests of other States in their exercise of the freedom of the high
seas'.[10]

While the doctrine of the abuse of rights has been recognised by the
World Court,[11] it is still somewhat controversial. Clearly, it can be
characterised as a general principle of law, and as such can be a source
of international law,[12] and it may be the genesis of a rule of customary
law, such as the rule enunciated in the *Trail Smelter Arbitration*.[13]
However, the operation of the doctrine is not free from difficulty. In
the words of one scholar:

> [t]here is no legal right, however well established, that could not,
> in some circumstances, be refused recognition on the ground that it
> had been abused. The doctrine of abuse of rights is therefore an
> instrument which . . . must be wielded with studied restraint.[14]

The assertion that the doctrine falls to be applied in a particular case
may be no more than a call for a more exact legal regime, and this may,
at one time at least, have been particularly true of the law relating to
pollution.

1.1 United Kingdom Common Law

At the municipal level within the UK, there has been only one decision
in which the common law on oil pollution has been at issue. In *Esso
Petroleum Co. Ltd* v. *Southport Corporation*[15] the beach at Southport
was damaged by oil jettisoned from an oil tanker within territorial waters.
A claim for damages alleging nuisance, trespass and negligence was
raised by Southport Corporation against the owners of the tanker. The
House of Lords took the view that it was unnecessary to consider whether
such a claim based on nuisance and trespass was competent, as the oil
had been discharged in order to save the lives of the crew. On the question
of negligence, it was held that the Corporation had not established the
allegations in their pleadings, and the action therefore failed. As a
precedent of common law, the case is of limited value.

It would appear that the English principle enunciated in *Rylands* v.
Fletcher,[16] and the corresponding Scottish principle, in *Kerr* v. *Earl of*

Orkney,[17] apply only to the escape of dangerous things from land and
that their extension to a discharge of oil by a ship either on the high
seas or within territorial waters is probably unwarranted.[18]

However, given the present delimitation of the UK continental shelf
for jurisdictional purposes,[19] which places the vast majority of the
North Sea oilfields appertaining to the UK within the Scottish sector,[20]
it is apposite to consider the position at common law in Scotland. The
decision in *Esso Petroleum Co. Ltd* v. *Southport Corporation* is not
binding in Scotland. Scots law does not attach the same meaning as
English law to the terms 'nuisance', 'trespass' and 'negligence', but
instead recognises a general principle that no one is entitled to do anything on his own property which will interfere with the natural rights
of others (*sic utere tuo ut alienum non laedas*). One particular aspect
of this principle is nuisance, for which liability in Scots law is strict.
While, normally, nuisance is a continuing infringement of another's
rights, there is authority for the view that one incident is sufficient
to constitute nuisance,[21] and so the discharge of oil which subsequently
fouls a beach or otherwise damages another person's property could in
Scots law amount to a nuisance.

That there are rules of customary international law and of common
law concerning pollution is clear. What is equally clear is that those
rules are too general or too skeletal or too fragmentary to cope with
the pollution problems that arise in practice, and these rules have been,
and are being, provided with flesh and made more precise by international agreements, and subsequent municipal legislation. There are
now in the order of a dozen major international agreements, not all of
which are yet in force. Notwithstanding that, however, the UK Government has implemented most of these conventions and, in many
instances, the implementing legislation, or at least part of it, has been
brought into force before the convention itself.[22]

1.2 Definition of 'Oil'

Pollution of the seas by oil, particularly as a result of the stranding of
a large tanker such as the *Amoco Cadiz*, or *Christos Bitas*, probably
attracts more press coverage than pollution from any other source, but
before dealing with the relevant legal framework, it is important to ask
what may seem to some a very simple question: what is oil? The
question becomes a little less simplistic when it is put in the form: how
is 'oil' defined in the various agreements and statutes on oil pollution?

An examination of the conventions and statutes reveals several
different definitions of 'oil'.

In the International Convention on Civil Liability for Oil Pollution Damage (CLC), 'oil' means 'any persistent oil such as crude oil, fuel oil, heavy diesel oil, lubricating oil and whale oil whether carried on board a ship as cargo or in the bunkers of such a ship'.[23] However, although that Convention is implemented by the Merchant Shipping (Oil Pollution) Act 1971, that Act uses the term 'persistent oil'[24] but fails to define it.[25] However, for the purposes of insurance against damage caused by oil pollution,[26] 'persistent oil' is defined in the Oil Pollution (Compulsory Insurance) Regulations 1977,[27] to mean any of the following:

(a)   hydrocarbon mineral oil whether crude or distilled, including crude coal tar and the oily residue of tank cleaning operations necessitated by the carriage of any such oils, but excluding those oils which consist wholly of distillate fractions of which more than 50 per cent by volume distill at 340° centigrade when tested by the 'American Society for Testing and Materials Specification D 86/67' in the case of oils derived from petroleum and at 350° centigrade in the case of oils derived from coal tar;

(b)   residual oil, consisting of mineral hydrocarbons comprising the residues of the process of distilling and/or refining crude petroleum and any mixture containing such residual oil;

(c)   whale oil.

That definition is much narrower than that in the CLC. However, there still remains the problem of what meaning is to be given to the term 'persistent oil' as it appears in other sections of the Act. It would be difficult to argue that the definition in the Regulations applies for all purposes, because the opening words of the Regulations are: 'for the purposes of Section 10(1) of the Act'. It is therefore suggested that the courts would look to the definition in the CLC in the absence of guidance in the Act.[28]

The International Convention for the Establishment of an International Fund for Compensation for Oil Pollution Damage of 1971 (Fund Convention) restricts the term 'oil' to 'persistent hydrocarbon mineral oils',[29] which is the general definition in the Merchant Shipping Act 1974.[30] However, for the purposes of determining who should make a contribution to the Fund, the word 'oil' is differently defined, but in this connection the definitions in the Fund Convention[31] and in the 1974 Act[32] are the same. Thus far, we have a definition in the CLC but none in the 1971 Act, except for the purposes of insurance. There is a different definition in the Fund Convention which is repeated in the 1974 Act. Although the possible difficulties arising from these separate definitions of 'oil' were pointed out by Mr Ronald King Murray MP in the Standing Committee Debate on the Merchant Shipping Bill,[33] he did not receive any satisfactory explanation. However, in the course of his observations, Mr Cranley Onslow, the Under-Secretary of State for Trade and Industry, said: 'As we are legislating to enact a Convention, it seems to me that we must enact what is said in the Convention.'[34] That point was not apparently regarded as important in the debates on the Merchant Shipping (Oil Pollution) Bill, in that no mention is made of a need to reflect the wording of the CLC.[35]

To compound the difficulties, there is a further definition of 'oil' in the International Convention for the Prevention of Pollution of the Sea by Oil 1954, as amended, where the term means 'crude oil, fuel oil, heavy diesel oil and lubricating oil'.[36] That general definition has remained since 1954, but the definition of 'heavy diesel oil' has altered.[37] Fortunately, the Prevention of Oil Pollution Act 1971 has the same definitions.[38] That Act[39] also implements the International Convention Relating to Intervention on the High Seas in Cases of Oil Pollution Casualties 1969 (Intervention Convention), which also uses the same definition.[40]

It would seem likely that these disparate definitions could give rise to problems. Since the definition in the CLC (which will probably be that applied for the purposes of the 1971 Act) is wider than that in the 1974 Act, there could be liability under the 1971 Act, but none on the Fund, for example for pollution damage caused by whale oil which is excluded from the definition in the 1974 Act. Likewise, although there could be civil liability under the 1971 Act in respect of whale oil, there would not be any criminal liability, for a deliberate discharge of 'whale oil' is not part of the definition in the Prevention of Oil Pollution Act. Perhaps few vessels carry whale oil in such bulk as to cause extensive damage in the event of a spill, but a close examination of the definitions show that there are other differences.[41] It remains to be seen how, if at all, the courts reconcile these differences.

2 ACCIDENTS

Accidents, such as the strandings of the *Torrey Canyon* and, more recently, the *Amoco Cadiz*, can result in enormous quantities of oil escaping into the sea.[42] However, the total amount of oil entering the sea annually as a result of accidents is small in comparison with other sources, amounting to about 8 per cent of the total, as can be seen from Table 1. None the less, it is accidents at sea that attract the greatest publicity and, at the same time, seem to prompt a substantial amount of governmental initiative; readers will doubtless recall the enormous press coverage devoted to the *Torrey Canyon* and *Amoco Cadiz* incidents, and the reactions of the UK and French Governments to these incidents.

It is clearly impossible to eliminate accidents completely, so efforts have been concentrated on ensuring that all ships are well constructed and well equipped, and that they are manned by competent crews.

2.1 Construction and Equipment

The basic international agreement is the 1954 Convention which emerged from the Inter-Governmental Maritime Consultative Organisation (IMCO). That Convention has been amended on three occasions, in 1962, 1969 and 1971, and is eventually to be superseded by the International Convention for the Prevention of Pollution from Ships 1973 (MARPOL). The 1954 Convention and the 1962 and 1969 Amendments are now in force, though the 1971 Amendments and the 1973 Convention are not.

The 1971 Amendments require that all tankers ordered after 1972 have their cargo tanks constructed and aligned in such a way that, given certain damage to the vessel, only a limited amount of oil can enter the sea. The formula for expressing this limitation is complex;[43] put briefly, the amount of oil that may enter the sea, the 'hypothetical outflow of oil' as it is called, is not to be in excess of a figure related to the tanker's deadweight, and must not in any event exceed 40,000 cubic metres. While these Amendments are not yet in force internationally, it appears that tankers are being designed and built so that their cargo tanks are constructed and arranged to comply with these standards. The enforcement of these standards is in the hands of the flag State and the port State. The State whose flag the tanker flies is obliged to require all its registered vessels to meet these construction standards; and to issue a certificate of compliance where they are met. Any State into whose ports or off-shore terminals the tanker comes may inspect the vessel and, if the tanker does not meet the standards, ultimately deny the tanker access.

Part II of the 1974 Act,[44] which implements these 1971 Amendments, is concerned with the design and construction of oil tankers,[45] the aim being to restrict the size of cargo tanks and thus reduce the amount of oil which may escape if tanks are breached. After the Act comes into force, all UK tankers, including those built before that date, will be required to comply with certain standards which will be laid down by the Secretary of State in 'oil tanker construction rules';[46] otherwise, they will not be allowed to proceed or attempt to proceed to sea.[47] Foreign tankers will require to comply with these standards[48] and have certificates to that effect,[49] failing which they may be refused entry to the UK or, if permitted entry, it will be subject to conditions imposed by the Secretary of State.[50]

In addition, these 'oil tanker construction rules' will replace two sets of statutory regulations[51] which are in force under the Prevention of Oil Pollution Act 1971,[52] which attempt to prevent or reduce the discharge of oil or oily mixtures into the sea, but only in so far as they relate to tankers.

The 'oil tanker construction rules' will implement the 1969 and 1971 Amendments to the 1954 Convention, thus restricting the amount of oil which may be discharged from tankers and placing a limit on the size of oil cargo tanks in tankers. Where a tanker has been ordered after 1 January 1972 or delivered after 1 January 1977, the wing tanks must not exceed 30,000 cubic metres and the centre tanks must not exceed 50,000 cubic metres.[53]

The powers in the 1974 Act[54] are wide enough also to implement MARPOL, which is to supersede the 1954 Convention and its Amendments. It reduces even further the acceptable level and rate of oil discharges,[55] but it also imposes a requirement on new oil tankers of more than 70,000 tons deadweight to have segregated ballast tanks.[56] An IMCO conference in February 1978 adopted Protocols to that Convention and to the International Convention for Safety of Life at Sea 1974 which further improve the construction of, and equipment in, tankers. New tankers of 20,000 tons or more must have segregated ballast tanks and existing tankers must have either segregated ballast tanks or clean ballast tanks or a crude oil washing system.[57] The 1979 Act contains provisions designed to give effect to these requirements.[58]

The 1978 Protocol also requires tankers of 20,000 tons or more to have an inert gas system to reduce the risk of explosion, and will supplement the existing rules about the prevention of fires[59] and the provision of fire appliances on board tankers.[60] The definitions of 'tanker' in these UK Rules[61] are not identical and they are narrower than the definition in the 1974 Act.[62]

Various other provisions are aimed at reducing the number of accidents. For example, in 1968, IMCO adopted recommendations about compulsory radar on ships of 1,600 gross tons or more and about compulsory echo-sounders on ships of 500 gross tons or more.[63] The recommendation about radar was implemented in the UK in the Merchant Shipping (Radar) Rules 1976,[64] but the other recommendation has not yet been implemented in the UK. The 1978 Protocol to MARPOL requires all ships between 1,600 and 10,000 tons to have one radar system, and all ships over 10,000 tons to have two such systems and duplicate steering gear.[65]

In December 1975, IMCO adopted a Resolution[66] entitled 'Procedure for the Control of Ships', by which States which are parties to the International Convention on the Safety of Life at Sea 1960 (now SOLAS 1974) and the International Convention on Load Lines 1966 were invited to adopt detention measures in respect of substandard ships. Although the Resolution attempted to lay down some tests by which a ship could be considered to be substandard, it was noted that a ship could not always be condemned solely by reference to a 'check-list', and that the person inspecting the ship would have to use his discretion. It was, however, resolved that should a ship be thought to be substandard, that fact could be brought to the attention of a State by a member of the crew, a professional organisation, a trade union or anyone with an interest in the safety of the ship, the crew and passengers. The 1978 Protocol to SOLAS 1974 tightens up the rules on inspection and certification,[67] and the 1979 Act will give effect to both the Convention and the Protocol.[68]

After a series of collisions in the English Channel, IMCO agreed, at the instigation of the UK Government, to a 'traffic separation' scheme which became compulsory in 1972 with the revision of the International Regulations for Preventing Collisions. The UK adopted the scheme on 1 September 1972.[69] The revised Collision Regulations also give preference to large ships which might have difficulty in manoeuvring, and they introduced safe speeds which depend on, and vary with, a number of factors such as visibility, traffic density, manoeuvrability and draught.[70]

It is all very well to make provision for the design and construction of vessels, and to institute traffic separation schemes, but all these will be of no avail unless adequate standards of crewing are also established. There are real difficulties in establishing such standards. As has been pointed out by one author, there is no guidance in the standards as to how they are to be achieved by the shipowners and there is no means by dictate of law alone to make 'a man a good and conscientious seafarer'.[71]

The existing international standards are couched in broad terms, requiring each State to 'undertake, each for its national ships, to maintain or, if it is necessary, to adopt, measures for the purpose of ensuring that, from the point of view of safety of life at sea, all ships shall be sufficiently manned'.[72] That broad obligation has to some extent been supplemented by more detailed rules.[73]

2.2. Standards for Crew

There are no special statutory provisions regulating the manning of oil tankers, so the requirements about training and competence of masters and crews are to be found in the general Merchant Shipping legislation. The present provision is s. 5 of the 1948 Act and regulations made thereunder.[74] New regulations may be made under the 1970 Act.[75]

However, much of the accidental oil pollution is the result of human error often caused by inadequately trained staff.[76] IMCO has been considering this problem for a number of years and in 1978 an International Convention on Standards of Training, Certification and Watchkeeping for Seafarers was drawn up. The International Labour Organisation has also been active in their attack on substandard ships and in 1976 the Merchant Shipping (Minimum Standards) Convention was drawn and supplemented by the Merchant Shipping (Improvement of Standards) Recommendation 1976. These are designed to regulate and improve safety standards, standards of competency, hours of work and training,[77] and if a ship fails to comply with the requirements, it may be delayed or detained, although not unreasonably.[78] As yet, there is no legislation in the UK on these issues but, despite that, the Merchant Navy Training Board provides instruction for those intending to work on oil tankers,[79] as do some shipping companies,[80] and the Department of Trade has encouraged all those who would benefit from such training to undergo it at the appropriate stage in their career or service in these ships.[81]

The high standard which the United Kingdom requires for the construction of its ships applies equally to its oil tankers, and it may not be too long before the particular 'oil tanker construction rules' are in force. It seems that this standard is considerably higher than in some other countries[82] and, for a while, criticism was directed against certain States, for example Liberia, Panama and Greece, as 'flags of convenience States',[83] since tankers flying these flags were involved in a large number of accidents.[84] More recently, under the influence of the United Kingdom Government, attention has been focused on substandard ships, no matter what flag they fly,[85] and the IMCO rules will permit more stringent examinations to be conducted by port authorities.

While well designed tankers are clearly essential, the crew may be the 'weak link in the chain'. Human error cannot be eliminated entirely,

but the presence on board a tanker of a competent crew can reduce it
to a minimum. In that connection, the 1978 International Convention
is of the utmost importance.[86]

However, although the crew are highly trained, there may be criticism
of the master for failing to call for aid or notify authorities until almost
the last minute. This criticism was made against the masters of the
Amoco Cadiz[87] and the *Christos Bitas.*[88] There may therefore be some
scope for regulating the exercise of the master's discretion in these
matters.

3 DELIBERATE DISCHARGES AND DISCHARGES FROM NORMAL OPERATIONS OF SHIPS

As has been said, deliberate discharges of oil and discharges from the
normal operations of ships are responsible for more oil entering the sea
than accidental spillages, accounting for about 38 per cent of all the
oil entering the sea.[89] These discharges come in the main from tank
washings which take place at sea after the cargo has been delivered. The
reason for the tank washing is that many ships are required to carry
ballast water in their fuel or cargo tanks in order to remain seaworthy.
In some cases, a ship will carry sea-water which is the equivalent of
between 30 and 40 per cent of its deadweight tonnage, and that water
must be discharged before a fresh cargo can be taken on. Some large
container ships have sea-water in their fuel tanks for the same reason.

Because the cargo or fuel tanks have been full of oil, the sea-water
becomes contaminated with the oily residue in the tank; that may be
as much as 1 per cent of the tank capacity and represent thousands of
tons of oil which may be discharged into the sea as the result of tank
washing. Most ships also have oily water in their bilges and this too has
to be discharged from time to time, although it must be admitted that
this is less of a problem.

As a general rule, there are no on-shore facilities which are adequate
to deal with such quantities of oily ballast and, in any event, to dispose
of these on-shore would be time-consuming. That being so, it is necessary
to reduce the amount of waste to a minimum. This could be done by
cleaning the ballast tanks while the ship is at the port of discharge, but
the problems of doing so have been mentioned. Another possibility is
for the ship to have segregated ballast tanks, i.e. certain tanks are
designed to carry only ballast, and sea-water in these tanks should not
therefore be contaminated by oil. This system does not seem to have
the support of the UK shipbuilding industry.[90] The third method is to
retain some of the cargo on board which then becomes deadfreight.

Although this reduces, if not eliminates, the discharge of oil at sea, it is a costly method, since the earning capacity of the vessel is reduced. The fourth method is to use the system known as Load-on-Top which was introduced by the major oil companies in 1964. Very basically, the system operates on the principle that since oil floats on water all the tank washings can be collected in one tank and the water (or most of it) drained, leaving only a small amount of oil and salt water. (The best explanation of the operation of Load-on-Top is to be found in 'Clean Seas Guide for Oil Tankers: The Operation of Load-on-Top' (LOT) published jointly by International Chamber of Shipping and Oil Companies International Marine Forum in 1973). The system is operated by more than 85 per cent of the world's tankers.[91]

The 1969 Amendments to the 1954 Convention set standards for discharges of oil or oily mixtures that can only be satisfied by operating Load-on-Top. In relation to tankers, the formula adopted is this: the vessel must be proceeding *en route*; the rate of discharge must not exceed 60 litres per mile; the total quantity of oil discharged must not exceed 1/15,000 of the vessel's total cargo-carrying capacity; the vessel must be more than 50 miles from land.[92] MARPOL maintains similar permissible oil discharge criteria as the 1969 Amendments, with three improvements: the discharge criteria are to apply to non-persistent oils;[93] the maximum quantity of oil or oily mixture that may be discharged during a ballast voyage has been reduced to 1/30,000 of the total cargo-carrying capacity;[94] and for certain areas no discharges are permissible, viz. the Mediterannean Sea, the Baltic Sea, the Black Sea, the Red Sea and the Gulf.[95] MARPOL, which is not yet in force, specifies that a tanker must be fitted with the equipment necessary to operate LOT, viz. an oil discharge monitoring and control system and a slop tank arrangement.[96] The Convention also requires States to provide ports with adequate facilities for receiving oily wastes from ships and tankers.[97]

The Prevention of Oil Pollution Act 1971 deals with discharges by UK registered ships and discharges into UK waters. It prohibits ships registered in the UK from discharging oil in any part of the sea outside the territorial waters of the UK except in certain circumstances. The Oil in Navigable Waters (Exceptions) Regulations[98] narrate the circumstances in which oil may be discharged, and these regulations replicate and give internal effect to the 1969 Amendments.

Contravention subjects both the owner and the master[99] to a fine not exceeding £50,000 on summary conviction and an unlimited fine on conviction on indictment,[100] unless it can be established that the discharge was necessary and reasonable in order to save the vessel or someone's life or to prevent damage to the vessel or cargo.[101] It is also a defence to show that the oil escaped as a result of damage to the vessel or a leakage (provided that neither the leakage nor any delay in discovery of it was due to lack of reasonable care), that, as soon as practicable after the damage or leakage occurred, all reasonable steps were taken to prevent, stop or reduce the escape.[102] Clearly the words 'practicable' and 'reasonable' will be the subject of considerable debate.[103]

The 1969 Amendments[104] and MARPOL[105] require that an oil record book be maintained for every tanker, recording, *inter alia*, discharges of oil at sea. Within the UK, the provisions of the Prevention of Oil Pollution Act are enforced by requiring the masters of tankers (and ships of 80 tons or more which use oil for fuel) to keep in their custody on board the vessel detailed records of, *inter alia*, the loading and discharging of cargo, the cleaning of tanks, deballasting operations and the disposal of ballast,[106] and also about the transfer to and from ships while they are within territorial waters.[107] The Secretary of State may appoint inspectors to ensure that the records are kept and kept accurately,[108] and the power of inspection may be extended to ships of countries which are parties to the Conventions.[109] Failure to keep accurate records may result in a fine.[110]

Any fines which are imposed under the Act may be enforced by distress or poinding and sale of the vessel, tackle, furniture and apparel; and any person who has incurred expenses or may incur expense in making good oil pollution damage may receive the whole or part of the fine to defray that expense.[111] The fines imposed, however, seem to be on the low side. In 1977, there were 58 convictions in respect of which the total fines imposed were only £76,676, and awards totalled £7,879, but in 1978, there were only 46 convictions in respect of which the total fines imposed were £34,167 and the awards amounted to £3,251.[112]

Apart from the mandatory oil record books, one might question how the discharge standards can be enforced. How can a spotter plane identify a rate of discharge in excess of 60 litres per mile? How can it be established that the total discharged exceeds 1/15,000 of the vessel's cargo capacity? Further, while LOT was highly praised during the 1960s, it has been noted of late that crew activities did not ensure as much success for the system as had been hoped.[113]

4 OFF-SHORE OPERATIONS

From the available statistics, oil pollution from off-shore operations is small by comparison with that from other sources, accounting for no more than 2.5 per cent of all the oil entering the sea annually.[114]

Only a few serious incidents have occurred and even then in only one — at Santa Barbara in California in 1969 — was there any significant on-shore pollution, but unusual geological features contributed to that incident. In the North Sea, over 1,200 wells had been drilled since the early 1960s without significant mishap, until April 1977 when there was a blow-out on the *Ekofisk Bravo* platform in the Norwegian sector of the North Sea, 175 miles south-west of Norway, but only 205 miles from the nearest UK coast. In that incident, an estimated 22,500 tons of oil escaped in the course of 8 days and the Commission appointed by the Norwegian Government came to the conclusion that the *Ekofisk* incident confirmed that 'the pollution hazard involved in uncontrolled blow-outs in the North Sea is exceptionally high'.[115]

A State has rights in the continental shelf only by virtue of a per-missive rule of international law, and, just as international law grants the rights, so it regulates how these rights are to be exercised, but only in the broadest terms. The Convention on the High Seas[116] and the Con-vention on the Continental Shelf,[117] both adopted in Geneva in 1958, set down general obligations concerning pollution: the former requires every coastal State to draw up regulations to prevent pollution of the sea by the discharge of oil from, *inter alia*, pipelines and off-shore installations; and the latter requires every coastal State to undertake in the 'safety zones', extending to 500 metres around off-shore installations appropriate measures for the protection of the living resources of the sea from harmful agents. Coupled with these conventional obligations, there is a general customary obligation, that is taken as extending to off-shore activities,[118] to the effect that a State must not exercise rights that it enjoys in such a way as to cause injury or damage to a neighbouring State.[119]

These general obligations have not been supplemented, like those on pollution from vessels, by specific international rules and standards, and the existing detailed regime of control is basically the result of municipal initiative.

Pollution from off-shore operations may be caused in several different ways, among which are blow-outs, collisions between ships and instal-lations, spills and leakages from storage terminals or pipelines, both accidentally and as the result of sabotage.[120] The United Kingdom legislation dealt with all of these situations even before the blow-out on *Ekofisk*, but since that time, the Secretary of State has required off-shore operators to submit their emergency plans for his scrutiny.[121]

The ownership of petroleum and natural gas was vested in the Crown by the Petroleum (Production) Act 1934[122] which also authorises the Secretary of State (at the Department of Energy) to grant licences for exploration and production.[123] Following the Geneva Convention on the Continental Shelf, the UK Parliament passed the Continental Shelf Act 1964 which extended the provisions of the 1934 Act to the UK off-shore areas. The areas of the North Sea which are within the jurisdiction of the UK are specified in statutory instruments.[124] Both the exploration and production licences incorporate Model Clauses which are set out in the Petroleum (Production) Regulations which have been modified to take account of changes in conditions and increase in expertise.[125] These Model Clauses cover such matter as: (a) working obligations; (b) maintenance of records; (c) supply of information to the Secretary of State; (d) consents to drill and abandon wells; (e) control of the flow of oil; (f) prevention of escapes and the reporting of escapes; (g) interference with fishing and navigation; (h) health, safety and welfare of personnel.

As has been said, the Department of Energy grants exploration and production licences, but under the Coast Protection Act 1949[126] the Department of Trade regulates the siting of installations. Consultations take place with various interested parties to ensure that the proposed operations will not be on major shipping lanes, nor on fishing grounds. In addition, to avoid collision or sabotage, the Secretary of State has created safety zones of 500 metres around fixed structures.[127] Furthermore, in 1975, the Minister of State for Defence indicated that arrangements had been made to protect installations in the UK sector against sabotage.[128]

A licensee, whether for exploration or production, must maintain all apparatus and appliances in good repair and condition and must execute all repairs in a proper workmanlike manner in accordance with 'good oil field practice'.[129] This in effect incorporates into all licences the standards laid down by the Institute of Petroleum's *Model Code for Safe Practice.*[130] Further, more particular obligations, imposed by licensees, may be supplemented by the Minister.[131] They must take all practicable steps to prevent discharge of oil[132] and must notify all escapes and discharges to the Secretary of State and to the Chief Inspector of HM Coastguard.[133] The operation of the licence must not interfere with navigation, fishing or conservation of the living resources of the sea.[134] All wells which have been abandoned must be satisfactorily plugged,[135] and petroleum obtained from off-shore operations is to be confined in accordance with 'good oil field practice' in tanks, pipes, pipelines etc. constructed for the purpose.[136]

The licensee's obligations under the Model Clauses are supplemented in two ways. The first relates to insurance,[137] in that certain additional obligations about construction, maintenance and operational procedure will be imposed on those who take out policies under the Offshore Pollution Liability Agreement (OPOL)[138] and in the Hull Clauses Form of insurance for installations or in the London Standard Drilling Barge Form. They are further supplemented by statute. The ultimate sanction for breach of one of the conditions of the licence is revocation of the licence[139] and, in the event of breach of one of the conditions of the insurance policy, the policy could be avoided. However, an inquiry which was set up to inquire into the collapse of the mobile jack-up platform *Sea Gem* in 1965 recommended that there should be a statutory code backed up by sanctions more effective than revocation of the licence, which admittedly has not proved necessary thus far.[140] The Mineral Workings (Offshore Installations) Act 1971 gave the Secretary of State powers to produce such a code by delegated legislation.[141] The Petroleum & Submarine Pipelines Act 1975, Part III of which came into force on 1 January 1976, requires that pipelines laid after that date be authorised by the Secretary of State, who may give directions about siting, capacity and design. To date, eleven sets of regulations have been made under the 1971 Act[142] and two under the 1975 Act.[143]

The Regulations under the 1971 Act are concerned mainly with protecting those who work on installations, but they are also aimed either directly or indirectly at preventing pollution. For example, particulars of each installation operating in the UK area must be entered in the Register of Offshore Installations maintained by the Department of Energy,[144] and each must obtain a Certificate of Fitness[145] which will be issued only after a major survey of the design and construction of the installation.[146] The Certificate may specify the areas into which the installation may go and the activities which it may undertake,[147] and failure to comply with these directives or any other regulation in respect of the installation may result in termination of the Certificate.[148] Installations are inspected annually to ensure that they remain safe and secure.[149]

Each installation has its own master who is responsible for the safety of the installation and personnel thereon and also for the maintenance of discipline.[150] A log book must be kept, recording, *inter alia*, details of all emergencies and action taken to deal therewith.[151] The implementation of these requirements is supervised by inspectors, whose powers of access and inspection are compulsory.[152] They may require that an

installation alter or cease operation[153] and they may report a manager
for prosecution.[154]

As has been shown by the *Ekofisk* incident, blow-outs can create
considerable problems. Part VIII of the Institute of Petroleum's *Model
Code of Safe Practice in the Petroleum Industries* is, as has been said,
effectively incorporated into the Model Clauses,[155] which provide that
each well must be fitted with a blow-out preventer (BOP) attached to
the well-casing and designed to cut off the well completely in the event
of an emergency. However, the blow-out on *Ekofisk* happened during
the fitting of a BOP because a down-hole safety valve (DHSV) had
not been properly locked into position. In such an event, there has to
be an emergency procedure for protecting human life, for stopping the
flow of oil and cleaning up what has already escaped. This emergency
procedure will also be brought into operation in the event of a fire,
which is the other major risk on board a rig or platform. The subordinate
legislation under the Mineral Workings (Offshore Installations) Act
1971 is aimed primarily at health, safety and welfare of personnel on
board these devices and is strictly observed.

It is difficult to envisage how the statutory provisions and delegated
legislation within the UK could further reduce the risk of oil entering
the sea from off-shore operations. The statistics demonstrate that the
amount of oil entering the sea from operational or deliberate discharges
in the course of off-shore operations is small,[156] but it has to be
acknowledged that a substantial amount of oil may escape as the result
of an accident. That said, attention must be focused on remedial measures,
as much for off-shore operations as for other sources of oil pollution.

5 REMEDIAL MEASURES

Although, as has been noted, there are various provisions designed to
reduce accidents to a minimum, it is clearly important in the event of a
spillage that effective remedial measures are taken immediately. When
the *Torrey Canyon* stranded in 1967, it was thought desirable to bomb
the vessel, but there was doubt about the legality of such a step,[157]
largely because the UK had acted against a foreign-registered vessel on
the high seas. In the wake of the *Torrey Canyon* incident, and prompted
by the UK Government, IMCO adopted the International Convention
relating to Intervention on the High Seas in Cases of Oil Pollution
Casualties in 1969[158] (Intervention Convention). Under this Convention
a State is authorised to take action on the high seas in the event of a
grave and imminent danger of pollution to its coastline or 'related
interests' following a maritime casualty.[159] Before acting, the endangered

State must undertake consultations, particularly with the flag State, and the action taken must be proportionate to the actual or threatened damage.[160]

5.1 Prevention of Oil Pollution Act 1971

The Prevention of Oil Pollution Act 1971 implements the Intervention Convention and permits the Secretary of State to take action in certain circumstances. The Act provides that where a ship, whether registered in the UK or not,[161] is involved in an accident and the Secretary of State is of the opinion that oil from the ship may or will cause 'pollution on a large scale' in the UK or its territorial waters, he may exercise certain powers if they are 'urgently needed'.[162] He may give directions to the owner, master or salvor or anyone else in charge of the ship or salvage operations about the movements of the ship, discharge of the cargo and salvage measures.[163] If these are ineffective, he has certain additional powers, including taking over the control of the vessel and even destroying it or part of it.[164] Any person who refuses to comply with such a direction or wilfully obstructs the exercise of these powers shall be guilty of an offence.[165] If the action taken by the Secretary of State was not reasonably necessary or the expense was inordinate compared with the benefit, any person who suffers loss as a result may seek compensation from the Secretary of State.[166]

The Act[167] uses the term 'accident', while the phrase in the Intervention Convention[168] is 'maritime casualty'. The latter phrase is sufficiently widely defined[169] to cover a deliberate discharge of oil, but the use of 'accident' in the Prevention of Oil Pollution Act is not. Thus, if a deliberate discharge would cause 'pollution on a large scale', could the Secretary of State exercise the powers conferred by s. 12? The term 'occurrence' in the Merchant Shipping (Oil Pollution) Act 1971 is preferable, as is 'maritime casualty' as defined in the Intervention Convention. Furthermore, before the Secretary of State can intervene, he must be certain the 'pollution on a large scale' is at least likely, but the Prevention of Oil Pollution Act does not give any guidance on what amounts to 'pollution on a large scale'. The Intervention Convention talks of 'major harmful consequences' which are defined to cover damage to the coastline and 'related interests', such as fishing, tourism, health, wildlife and marine resources,[170] but the Act does not make it clear whether these interests are included. If, for example, there is a collision as the result of which oil escapes from a vessel and is unlikely to come ashore, but may cause damage to birds and marine resources, could the Secretary of State assume the powers conferred by s. 12? It is possible

that since the Act is unclear on this point, recourse might be had to
the Intervention Convention.[171] Art. III of the Convention deals with
the conditions which apply to the exercise of the powers conferred in
Art. I but, in this connection, the Act departs significantly and
importantly from the Convention. The Convention provides that before
taking any action, the State authority should, *inter alia*, 'proceed to
consultations with other States affected by the maritime casualty,
particularly with the flag State or States', and it must take account of
the views expressed. There is no corresponding provision in the Act. It
is to be assumed that the Secretary of State would consult those with
an interest in the vessel or other States suffering pollution from the
same source, but he need not and, in this respect, the Act is not in
conformity with international law. There is undoubtedly a presumption
that Parliament does not intend to legislate in a manner which is
inconsistent with international law,[172] but it is submitted that it would
be extremely difficult to read into the 1971 Act an *obligation* to consult,
because there does not seem to be any ambiguity in the Act which
would permit this step. However, the Department of Trade's Central
Unit on Environmental Pollution has pointed out that the UK would
need to take account of the interests of its neighbours where they were
affected by the oil.[173]

The UK legislation encourages the use of remedial measures in every
case, especially in major incidents. The Merchant Shipping (Oil Pollution)
Act 1971 makes the owner liable for any measures reasonably taken after
an escape or discharge to prevent or reduce damage in the area of the UK,
and also for any damage caused in the UK by such measures.[174] If the
owner takes remedial measures, he may recover the cost thereof from
any limitation fund set up under the CLC and s.5 of the Act, and if he
requires to claim against the International Fund set up under the Fund
Convention and the 1974 Act (although the liability of the Fund may
be reduced or eliminated if the person claiming has intended to cause
damage or been negligent),[175] that does not apply to expenses reasonably incurred or sacrifices reasonably made to prevent or minimise oil
pollution damage.[176]

The last method by which remedial measures are encouraged is to be
found in the sections dealing with defences to contraventions of ss. 1,
2 and 3 of the Prevention of Oil Pollution Act, which deal with discharges
from ships and pipelines. It is a defence to a charge under these sections
that as soon as practicable after the discharge or escape occurred, all
reasonable steps were taken to prevent, stop or reduce the escape of oil
or the mixture.[177] For the purposes of s. 12, '"oil" means "oil of any

description"'[178] and this includes 'oil' as defined in the Intervention Convention[179] and 'oil' as defined in the 1973 Protocol to that Convention.[180]

If a spillage of oil threatens coastlines or ports, local authorities and port authorities are empowered to take steps to deal with such spillages,[181] but while these and the above-mentioned measures encourage persons to take remedial measures, no one is under a statutory obligation either to take such measures or to draw up contingency plans.[182] In a White Paper issued in January 1969,[183] the Government undertook to introduce a Bill requiring local authorities to draw up plans to deal with oil spills, but no such Bill was introduced. All that exists at present is a guidance circular issued in 1968 by the Ministry of Housing and Local Government,[184] the Welsh Office[185] and the Scottish Development Department.[186] These requested local authorities to draw up contingency plans pending legislation. The need for plans was indicated by the Department of Trade's CUEP[187] and their absence was criticised in the Report of the Select Committee on Science and Technology on the *Eleni V* incident.[188] The Department of Trade have recommended that a small unit be set up within its Marine Division and that it should have responsibility for planning and emergency operations,[189] but it recommended that the present division of responsibility between that Department and local authorities should be maintained[190] and rejected the notion of a body solely responsible for oil pollution contingency arrangements.[191] However, it did recommend that the various contingency plans should be submitted to the Government and kept under review.[192] In November 1978, it was announced in Parliament that a Marine Pollution Control Unit had been set up as part of the Marine Division of the Department of Trade with the following main aims:[193]

i to ensure that the arrangements for using the resources available for dealing with oil pollution at sea are as effective as they can be;
ii to develop a national plan, including measures to deal with potential marine pollutants such as chemicals and other dangerous cargoes;
iii to relate these plans to those of neighbouring countries so as to provide as much natural support as possible;
iv to take charge of operations at sea in the event of a marine pollution emergency in British waters.

As far as off-shore operations are concerned, special arrangements have been made in an attempt to ensure speedy and effective remedial measures. The responsibility for emergency procedures lies in the first instance with the operator who must have a contingency plan approved by the Government. The United Kingdom Offshore Operators' Association (UKOOA) also gives guidance on this subject and there is liason with the Department of Energy. The operator has insurance cover for any damage, and compensation is also available under OPOL.

The Offshore Installations (Emergency Procedures) Regulations[194] require operators to have contingency plans, but, in the light of *Ekofisk*, these plans were called in by the Government and in March 1978, the Departments of Trade and Energy jointly produced a pamphlet entitled *Oil Spill Clean-Up Arrangements in the Offshore Oilfields* which contained 'Notes on the Principal Elements of an Oil Spill Contingency Plan'. In August 1978 the oil companies themselves announced details of their current plans.[195]

Operators may draw on the resources of the UKOOA which has produced a booklet entitled *Offshore Safety, Emergency Techniques and Environmental Protection*, and through its Clean Seas and Emergency Services Committee has created stocks of dispersants and spraying apparatus. One of the defects highlighted by Paul 'Red' Adair, whose team was instructed to deal with the blow-out on *Ekofisk*, was the paucity of equipment, especially fireships. The operators were therefore encouraged to form 'Sector Clubs', which they have done, and they now have a substantial number of fireships and emergency vessels.[196]

6 CIVIL LIABILITY AND COMPENSATION

The stranding of the *Torrey Canyon* in March 1967[197] on the Seven Stones, off the west coast of England, marked a turning-point in the law relating to pollution, not least in the area of liability. It served to demonstrate the inability of existing UK law to deal with the problems:

(i)   whether the English courts had jurisdiction to try an action for compensation. The vessel was owned by the Barracuda Tanker Company of Bermuda which was associated with the Union Oil Company of Los Angeles and, for taxation purposes, was registered in Liberia. At the time, the ship was on lease to the Union Oil Company, but had been chartered by British Petroleum Co. Ltd for a voyage from the Persian Gulf to Milford Haven. To complete the international flavour, the master and crew were Italian. The stranding occurred outside UK jurisdictional limits.

(ii) whether there was any basis in substantive law for such an action.

The only previous authority, *Esso Petroleum Co. Ltd* v. *Southport Corporation*,[198] was somewhat inconclusive.

(iii)   whether, even if jurisdiction could be established and a ground of action established, the owner might nevertheless be able to limit his liability. In the UK the shipowner (and various other people) can limit his liability if it can be demonstrated that the damage occurred 'without actual factual fault or privity' on his part; the compensation fund is calculated with reference to the tonnage of the vessel.[199]

(iv)   whether any judgment could have been enforced, given the cosmopolitan nature of the vessel.[200]

The *Torrey Canyon* stranding and the resultant pollution indicated a need for international accord on the problems mentioned above. That accord took various forms of which two international conventions are of particular importance, viz. the CLC adopted in 1969 and the Fund Convention drawn up in 1971. These have been enacted by the United Kingdom Government in the form of the 1971 Act[201] and the 1974 Act.[202]

6.1 Jurisdiction

The CLC requires that each contracting State ensures that its courts possess the necessary jurisdiction to entertain claims for compensation and, for that purpose, the vessels registered in each contracting State are deemed to be subject to the jurisdiction of every other contracting State in which pollution damage occurs.[203]

The 1971 Act provides[204] that claims shall be dealt with as admiralty claims, and that every State which is a party to the CLC is deemed to have submitted to the jurisdiction of the appropriate UK court in respect of any claims under s.1.[205] However, the UK courts will not entertain any action in respect of damage outside the UK, if no preventive measures have been taken in the UK, and the damage has been caused or expenses have been incurred in taking remedial or preventive measures[206] in the area of a Convention country. An action under s.1 must be commenced within three years of damage being caused or within six years of the occurrence or the first occurrence, whichever is sooner.[207]

6.2 The Basis of the Claim

The CLC provides[208] that the owner[209] of a ship[210] shall be liable for any pollution damage[211] caused by oil[212] which has escaped or been discharged from a ship at the time of an incident,[213] provided the damage was caused on the territory including the territorial sea of a contracting State.[214] Liability is strict and the owner is liable to 'any individual or partnership or any public or private body whether corporate or not including a State or any of its constituent sub-divisions'.[215] Where oil has escaped or been discharged from two or more ships and pollution damage occurs, the owners are jointly and severally liable if it is not reasonably clear which damage was caused by each ship.[216]

There are, however, various defences available, and the owner will not be liable if he can demonstrate that the damage was caused by any of the following, viz.:

(i)   an act of war, hostilities, civil war, insurrection or a natural phenomenon of an exceptional, inevitable and irresistible character;[217]

(ii)   an act or omission done with intent to cause damage by a third party.[218] (If the owner can prove that the pollution damage was caused in whole or in part by such an act or by the negligence of the person who suffered the damage, he may be exonerated in whole or in part from his liability to that person.)[219]

(iii) by the negligence or other wrongful act of any government or other authority responsible for the maintenance of lights and other navigational aids in the exercise of that function.[220]

The owner of a ship registered in a contracting State and carrying more than 2,000 tons of oil in bulk as cargo is required to maintain insurance or other financial security to cover potential liability.[221] Art. VII(1) gives two examples of acceptable financial backing, viz. a bank guarantee and a certificate delivered by an international compensation fund. Each ship will be issued with a certificate to the effect that such financial backing is in force;[222] that certificate must be carried by the ship and a copy deposited with the authorities who keep the ship's registry.[223]

The 1971 Act[224] imposes strict liability on the owner[225] of a ship[226] which is carrying a cargo[227] of persistent oil in bulk. S. 1 provides that where, as the result of any occurrence,[228] any such oil is discharged or escapes from the ship, the owner will be liable (a) for any damage[229] caused by contamination in the area of the UK[230] or of any Convention country;[231] (b) for the cost of any reasonable measures taken by the owner to reduce or prevent such damage; and (c) for any damage caused by taking such measures.[232] If there is an escape or discharge from two or more vessels and it is not possible to apportion liability, liability is joint.[233]

If the owner is not liable under s. 1, he is not liable at all,[234] and s. 3 exempts the owner's employees and agents, as well as any salvor acting with the owner's consent from liability under s. 1. The Act here departs from the CLC which does not exempt salvors from liability.[235] In the parliamentary debates on the Bill, the Under-Secretary of State for Trade and Industry gave, as the reason for the exemption, the difficulty which salvors might experience in obtaining insurance.[236] The reasoning suggests that others might seek exemption from liability on the ground that they find it difficult to obtain insurance, which is difficult to accept. However, unlike the owner, his employees, agents and the salvor would not escape liability at common law, although the owner would not be vicariously liable for their actions.[237]

In that connection, however, two other sections of the Act should be noted. The first is s.7, which provides that where the owner incurs a liability under s.1 and another person incurs liability other than under s.1 (for example the master or crew member in the above example), then subject to two conditions, no proceedings shall be taken against that other person; or, if proceedings have been commenced, the other person shall be liable only for costs (expenses). The conditions are (a) that the owner has been found entitled to limit his liability under the 1971 Act[238] and (b) that the other person is entitled to limit his liability under the 1958 Act.[239] The other is s.15, which provides that where oil has escaped or been discharged and a person takes steps to prevent or reduce the damage, then anyone who would incur, or, but for these measures, would have incurred, a liability other than under s.1, will be liable for the cost of these remedial measures. This is intended to assist, for example, local authorities who require to take steps to avert damage from oil pollution. Although the section says that such a person may recover the cost of remedial measures, it is not restricted to that cost if more substantial losses have been incurred.[240]

The owner may escape liability under s.1 if he can establish one of three defences,[241] which are similar to those available under the CLC,[242] but not identical.[243] In addition, the owner may plead that there was 'contributory negligence'[244] on the part of the claimant and so escape, or partially escape, liability.[245]

6.3 Limitation of Liability

The CLC entitles the owner of a ship from which the escape or discharge occurred to limit his liability[246] in respect of any incident to 2,000 francs[247] for each ton of the ship's tonnage[248] up to a maximum of

210 million francs, provided the incident occurred without actual fault or privity on his part. The construction of the phrase 'without actual fault or privity' has given rise to problems with which maritime lawyers are familiar,[249] but if the incident was caused by the owner's actual fault or privity, liability is unlimited.[250] If the owner wishes to avail himself of these limits, he must create a fund representing the total sum for which he may be liable under Art. III,[251] but those supplying the financial backing under Art. VII must also create such a fund,[252] because a person suffering pollution damage may sue either the ship-owner or the insurer.[253] The fund, once established, must be lodged with the court or other competent authority of any of the contracting States in which an action is brought under Art. IX.[254]

Liability under the 1971 Act is strict, but the owner may limit his liability thereunder[255] if he can show that the discharge or escape occurred without his actual fault or privity.[256] The owner must apply to the court for limitation,[257] but if it decides in his favour, then his aggregate liability under the section shall not exceed 2,000 gold francs[258] per ton of the ship's tonnage[259] or 210 million gold francs, whichever is less.[260] From time to time, the Secretary of State may specify the sterling equivalents of these figures.[261] If the ship is registered in a country which is not a party to the CLC, but is a party to the International Convention on the Limitation of Liability of Owners of Seagoing Ships 1957, the limits set out in the CLC are replaced by the lower limits set out in the 1957 Convention.[262]

If the court holds that the owner may limit his liability, it shall direct him to pay the total amount of his liability into court,[263] thus creating a fund which will then be apportioned and distributed among the claimants.[264] By creating such a fund, the owner is not affected by any subsequent changes in the value of sterling.[265] The other benefits of consignation are (a) the release of any arrested property or any security given to release the property or prevent its arrest[266] and (b) that no judgment (except for costs) shall be enforced in respect of an s.1 claim.[267] Consignation of the fund excludes further diligence (in Scotland)[268] and execution (in England)[269] against the owner's assets.[270] If the owner (or certain others, for example the insurer)[271] has made a payment to account before the action is raised, he shall be credited therewith in the eventual distribution.[272] Likewise, if a person has made a reasonable sacrifice voluntarily in the hope of preventing or reducing the damage, or has taken some other reasonable measure, he shall be credited with the cost of these also.[273]

6.4 Enforcement

In terms of the CLC, as long as the defendant is given reasonable notice of the claim and a fair opportunity to present his case, all contracting States must recognise and enforce the judgment,[274] a provision that is given effect in the 1971 Act by application of the Foreign Judgments (Reciprocal Enforcement) Act 1933.[275]

Together the CLC and the 1971 Act have gone a long way to resolve the major legal difficulties that confronted potential claimants at the time of *Torrey Canyon*. None the less, there remain several problems, not the least of which is the question of whether there are adequate funds available to meet claims in full. This question has been approached in two ways: by requiring shipowners to carry insurance and by establishing an international fund.

7 INSURANCE

The CLC provides[276] for compulsory insurance up to the limit of liability for all ships registered in a contracting State which carry more than 2,000 tons of persistent oil in bulk as cargo. Such ships will be issued with, and must carry on board, a certificate attesting that such insurance or other financial security is in force. The claimant is entitled to go directly against the insurer.

Similar provisions are found in the 1971 Act.[277] If a contract of insurance complies with the Act, the Secretary of State must issue a certificate to that effect,[278] but he may refuse to do so if there is doubt about the financial stability of the insurer.[279] He may cancel the certificate if it is established in legal proceedings that the insurance is or may be treated as invalid.[280] Concern has been expressed in the United States that some contracting States might issue certificates even although the financial backing was non-existent. In that event, the US, after consultation with the appropriate government, might refuse to recognise the certificate and require the ship to have a certificate issued by the US authorities before it entered the ports there.[281] Whether the UK Government would do the same remains to be seen.

It is clear that no ship will be allowed to enter the UK or to use any terminal in territorial waters unless insurance is in force and the ship carries a certificate to that effect.[282] If it does enter, it may be detained.[283] Failure to have the required insurance can result in the owner or master[284] being fined up to £35,000 on summary conviction and an unlimited fine on conviction on indictment.[285] If the ship does not carry, or the master fails to produce, evidence of insurance, the master will be liable to a fine not exceeding £400 on summary conviction.[286]

If a contract of insurance is in force, a person who has a claim for
s.1 damage may enforce it against the insurer rather than the owner.[287]
The insurer may limit his liability and may do so whether or not the
owner can do so.[288] The same defences are open to him, but, in addition,
he may plead that the discharge or escape was caused by wilful misconduct on the part of the owner.[289] Thus, it would seem that the
insurer would have a defence if the owner were convicted under ss.1 or
2 of the Prevention of Oil Pollution Act 1971,[290] or if the ship was
deliberately scuttled.[291]

The advantage of this arrangement is that a person suffering oil
pollution damage has a claim against a body whose financial standing
is probably higher than that of the owner, but against that, the insurer
will not be liable for the owner's wilful misconduct, and he can limit
his liability in circumstances in which the owner's liability would be
unlimited. Nevertheless, it might be advantageous to obtain a settlement
in part from the insurer and attempt to recover the balance from the
International Fund under the 1974 Act.[292]

As has been said, the person who suffers loss may not receive any
compensation from the owner or his insurer or be unable to recover the
full amount of his loss.

In November 1971 IMCO held a conference at Brussels to consider a
supplementary convention to the CLC. The result was the Fund Convention, the object of which was twofold; first, to provide compensation
for victims of oil pollution damage who have been unable to obtain a
complete remedy from the shipowner under the CLC, because no
liability arises under that Convention, or because the shipowner or his
insurer are insolvent, or because the shipowner has been able to limit
his liability; and, second, to indemnify the shipowner or the insurer for
the increased financial burden imposed by the CLC.

The International Fund is created by a levy on oil imported into, or
received by, the contracting States,[293] and the obligation rests on oil
companies, rather than governments. As has been observed, 'the
philosophy underlying this approach is that those who, in earning
substantial sums of money, expose others to the risk of large-scale
pollution should play some part in paying for any damage which
occurs'.[294]

The Fund Convention is given internal effect in the UK by the 1974
Act. The Secretary of State may require those in the UK who are liable
to contribute to find security for their contributions;[295] and to ensure
that those who are liable to contribute actually do so, the Secretary of
State has power to obtain information from 'any person engaged in

producing, treating, distributing or transporting oil'.[296]

8 COMPENSATION

S.4 of the 1974 Act deals with compensating those who have been unable to obtain any compensation or full compensation under the 1971 Act,[297] but there are certain circumstances in which there will not be any liability on the Fund. The first of these is where the claimant cannot identify the ship which caused the pollution.[298] (In these circumstances the Advisory Committee on Oil Pollution of the Sea (ACOPS) have suggested that there ought to be a Government Fund to compensate the victims.[299]) The second is where the ship which caused the pollution was a warship or other Government ship on non-commercial service.[300] In addition, the Fund will not make reparation if the damage 'resulted from an act of war, hostilities, civil war or insurrection',[301] which approximates to the common law notion of Act of God.[302] This repeats in part one of the defences in the 1971 Act,[303] but although the Act excludes damage resulting from 'an exceptional, inevitable and irresistible natural phenomenon'; the Fund will be liable in such circumstances. There will also be liability on the Fund if the escape or discharge (a) 'was due wholly to anything done or left undone by another person, (not being a servant or agent of the owner), with intent to do damage' or (b) 'was due wholly to the negligence or wrongful act of a government or other authority in exercising its function of maintaining lights or other navigational aids for the maintenance of which it was responsible'. It seems odd that, if the ship which caused the damage is a Government ship, the Fund is not liable, but, if the damage was caused by the negligence of the Government, the Fund will be liable.

9 INDEMNIFICATION

Under s.5, which deals with indemnification, the owner or insurer may be relieved of a portion of his liability under the 1971 Act,[304] unless the damage resulted from the owner's misconduct.[305] The Fund may escape liability completely or partly if, as a result of the actual fault or privity of the owner, the ship did not comply with certain requirements laid down by the Secretary of State and the damage was caused wholly or partly by that non-compliance.[306] The Secretary of State has now specified these requirements, which are similar to those in the Fund Convention.[307]

A UK court will not entertain an action against the Fund unless it is raised within three years of the date when a claim against the Fund arose, or a third-party notice of an action against the owner or his insurer is given to the Fund within that period.[308] In addition, an action will not be entertained if it is raised later than six years from the date of the occurrence which caused the discharge or escape.[309] These periods are not, however, absolute in that the 1974 Act provides that an owner or his insurer's rights to indemnity will not be extinguished before the expiry of six months from the time he was first aware that an action under the 1971 Act (or similar provision) had been raised against him.[310] That period is so short that the owner or insurer might be forced to make a provisional claim against the Fund, if the action under the 1971 Act has been not concluded. The wording of ss. 7 and 8 would appear to enable an action to be raised against the Fund, notwithstanding that an action against the owner or insurer is pending, but that is difficult to reconcile with the statement in s.4(1): 'The Fund shall be liable for pollution damage in the United Kingdom if the person suffering the damage has been unable to obtain full compensation under s.1 of the 1971 Act . . .' That would appear to contemplate an action being raised first against the owner or insurer and, only when the result of that is known, would it be competent to go against the Fund. It remains to be seen how these provisions will be reconciled.

Just as under the 1971 Act,[311] a claim against the Fund comes within admiralty jurisdiction[312] and any judgment is enforceable under the 1933 Act.[313]

Even before the adoption of the CLC and Fund Convention, tanker-owners and those transporting oil by sea had voluntarily entered into agreements providing a scheme for liability for both clean-up costs and oil pollution damage and a fund for compensation: these agreements are the Tanker Owners Voluntary Agreement concerning Liability for Oil Pollution (TOVALOP) and the Interim Supplement to Tanker Liability for Oil Pollution (CRISTAL).[314]

10 OFF-SHORE OPERATIONS (LIABILITY)

The discussion so far has dealt only with tankers, but similar problems can arise out of an occurrence such as the blow-out on *Ekofisk Bravo*; for example, questions of jurisdiction and actionability in respect of any damage caused by the oil or expense incurred to avert such damage. A full discussion of these is appropriate to a text on Conflict of Law, but some of the problems will be outlined here.

If some of the oil had come ashore in a part of the United Kingdom, that would have raised two important points: whether there had been a delict or tort and whether the courts had jurisdiction. On the assumption that there had been an actionable wrong (and that would be a matter to be determined by law of Scotland or England as the *lex fori*), the Scottish courts would assume jurisdiction if damage was caused in Scotland, regardless of where the breach of duty took place,[315] and it would seem that the English courts would adopt the same approach.[316]

If the rig or platform is in the UK sector of the continental shelf and it could be established that there had been a breach of a provision of the Mineral Workings (Offshore Installations) Act 1971 or any regulation thereunder, that would be actionable in the UK courts[317] if the breach resulted in personal injury or death.[318] Such an action would probably be available only if the injury occurred in a designated area or within territorial waters or on land.

If no oil actually came ashore, although an action might be available under the 1971 Act, what is most likely to happen is that the local authority or other body would take steps to prevent oil pollution damage and the question would then be whether they could recover the cost of their preventive measures. If such a body took action in respect of a spill from a ship, they would be able to obtain compensation under s.15 of the 1971 Act, but there is no statutory provision to deal with a spillage from an off-shore installation. That being so, one must have recourse to the common law. There seems little doubt that one may take reasonable steps to protect one's own property and that of others. If that involves expense, there does not seem to be any good reason why that expense cannot be recovered.

As for ships, an international agreement (the London Convention on Civil Liability for Oil Pollution Damage from Offshore Operations 1976) and an interim voluntary agreement among oil companies (OPOL) have been adopted for off-shore extractive operations.[319]

11 CONCLUSIONS

It would have been impossible in the space of one chapter both to set out all the relevant law (international and UK) and to discuss issues at any meaningful level. Of necessity, this chapter has been painted with a broad brush; but it is hoped that, at the same time, some of the major features in the picture can be identified. Many of the issues touched upon here will be taken up in later chapters.

To draw any conclusions from such a general study is somewhat dangerous. It is clear that there is now no dearth of legal rules relating to pollution. That, in itself, does not make the position satisfactory. Some of the international agreements are not yet in force, for example the 1971 Amendments to the 1954 Convention, MARPOL and the 1976

London Convention. Some States are not parties to some of the international agreements, while others, who are, have not as yet ratified them. This leads to a situation where, for example, 15 per cent of the world's tankers, which clearly are registered in States which are not bound by the 1969 Amendments to the 1954 Convention, and therefore are not obliged to operate LOT, contrive to cause 529,000 tonnes of oil to enter the sea annually;[320] this is five times the amount of oil entering the sea from those 85 per cent of tankers which do operate LOT, and more than twice the amount of oil that escaped from the *Amoco Cadiz*. Further, one shudders to think of the plight of anyone suffering pollution damage caused by a vessel registered in a State which is not a party to the CLC.

There are also, and possibly inevitably, *lacunae* and ambiguities in the international agreements and in the implementing legislation.[321] The varying definitions of 'oil' have already been discussed, and these could yet pose real problems. One problem of considerable magnitude that remains is that it may not be possible to identify the vessel that caused pollution damage. No recovery would then be possible under the CLC and Fund Convention and the two UK implementing statutes. In that connection, ACOPS has unsuccessfully attempted to persuade the UK Government to establish a fund to give compensation in such a case.[322] There is considerable force in their argument, because a survey of oil spills round the UK coast which they conducted demonstrated that in 308 out of a total of 432 incidents, the source could not be identified,[323] despite the existence of fairly sophisticated methods, 'tagging' and 'finger-printing', which can identify the source of an oil spill.[324]

The success, or otherwise, of all the international agreements turns on enforcement.[325] IMCO have recently stated that attention will now be paid to the implementation and enforcement of existing provisions rather than the adoption of new measures.[326] Despite the fact that the conventions provide for mutual enforceability by allowing each contracting State to enforce the relevant rules against the vessels registered in other contracting States, there remain problems. It may be that some provisions are unenforceable at law. In the absence of an accurate record kept on a tanker, how are the provisions of the 1969 Amendments to the 1954 Convention to be enforced? How can one detect at sea a discharge rate in excess of 60 litres per mile, and a total discharge of more than 1/15,000 of the vessel's capacity? The sanction here may not be legal, but may rather be economic, for LOT has been demonstrated to make good economic sense.[327]

It is all to easy to say that oil companies and governments could and should do more to combat oil pollution, and there is force in that argument. Yet the record of the oil companies is not as black as it is often painted. For whatever motive, it was the oil companies that introduced Load-on-Top in 1964, and established voluntary compensation schemes in TOVALOP, CRISTAL and OPOL. So far as the UK Government is concerned, despite criticisms of, for example, the Government's handling of the *Eleni V.* incident[328] and the absence of an international fund for compensation for pollution where the vessel cannot be identified, its record of giving internal effect to the international agreements is very good. ACOPS, which has not been slow to criticise Government activity, has observed that in 1978, it is probable that more parliamentary time was devoted to discussion of marine pollution than in any year since the stranding of *Torrey Canyon.*[329] Those who insist on using petrol in their cars, oil in their central heating and products derived from hydrocarbons must accept the inevitability of, and some responsibility for, oil pollution.

NOTES

1. The table is derived from information contained in IMCO Report of Study No. VI submitted by the United Kingdom: 'The Environmental and Financial Consequences of Oil Pollution from Ships', Appendix 1, 'Discharges of Oil into the Oceans', as interpreted by Abecassis, *Oil Pollution from Ships*, 4, and Keates, 'International Law and Oil Pollution of the Seas', address at the I.B.A. Committee on the Environment, Vancouver 1974. It must be appreciated that while these statistics give a broad overview of the problem, they should not be treated as definitive: the US National Academy of Science, 'Petroleum in the Marine Environment – Inputs', 1975.

2. The text of this Act is to be found in (1970) 9 I.L.M. 543; the Canadian justification for the Act is to be found in (1970) 9 I.L.M. 607.

3. For a full discussion of these issues see Birnie, Chapter 3 of this book.

4. On the nature of an international custom, see Brownlie, *Principles of Public International Law*, 2nd edn, 4-15; D'Amoto, *The Concept of Custom in International Law*.

5. (1940) 3 R.I.A.A. 1905.

6. See, for example, Fleischer, 'Pollution from Seaborne Sources' in Churchill, Simmonds and Welch, *New Directions in the Law of the Sea*, Vol. III, 78 at 80-2.

7. [1949] I.C.J. Rep.4 at 22.

8. [1956] 1 I.L.C. Yearbook 110.

9. Convention on the High Seas, 1958, Preamble, Arts. 2, 24 and 25.

10. Ibid., Art. 2.

11. *Certain German Interest in Polish Upper Silesia* (1926) P.C.I.J., Ser.A., No. 7, 30; *Free Zones Case* (1930) P.C.I.J., Ser.A., No. 24, 12.

12. Cheng, *General Principles of Law*, 121-36; Lauterpacht, *The Function of Law in the International Community*, 286-306.

13. Brownlie, *Principles*, 432.
14. Lauterpacht, *Development of International Law by the International Court*, 164.
15. [1956] A.C. 218.
16. (1866) L.R. 3 H.L. 330.
17. (1857) 20 D. 298.
18. Ingram, 'Oil Pollution – Rylands v Fletcher', (1971) 121 N.L.J. 183.
19. The Continental Shelf (Jurisdiction) Order 1968, S.I. 1968/892, as amended by S.I. 1971/721, 1974/1490, 1975/1708, 1976/1517, 1978/454.
20. Grant, 'Oil and Gas' in Grant (ed.), *Independence and Devolution: The Legal Implications for Scotland*, 88.
21. *Slater* v. *A. & J. McLellan* 1924 S.C. 854.
22. For example, the Merchant Shipping (Oil Pollution) Act 1971 came into force in part on 9 September 1971 (S.I. 1971/1423) and the remainder on 19 June 1975 (S.I. 1975/867) when the CLC came into force. The Prevention of Oil Pollution Act 1971 came into force on 1 September 1973. The 1969 Intervention Convention came into force on 6 May 1975, but to date only the 1969 Amendments to the 1954 Convention are in force. They are operative from 20 January 1978. The Merchant Shipping Act 1974 came into force in part on 1 November 1974 (S.I. 1974/1792) and further parts on 19 June 1975 (S.I. 1975/866), 16 October 1978 (S.I. 1978/1466) and 1 August 1979 (S.I. 1979/808). The Dumping at Sea Act 1974 came into force on 27 June 1974 but the two Conventions did not come into force until 30 July 1975 (Oslo) and 17 December 1975 (London).
23. Art. 1(5). A UK Government inter-departmental group, the Group on Liability and Compensation, has recommended that the definition be extended to non-persistent oils and has suggested that this be considered by the Legal Committee of IMCO. See 'Liability and Compensation for Marine Oil Damage' (the Department of Trade, February 1979), para. 24. It also recommended extending the definition to cover oil in the bunkers of unladen ships (para. 29).
24. Ss. 1-3.
25. Ss. 19-20 are the definition sections.
26. S. 10(1).
27. S.I. 1977/85, Reg. 3.
28. *Salomon* v. *Customs & Excise Commissioners* [1967] 2 Q.B. 116.
29. Art. 1(2).
30. S. 1(3).
31. Art. 1(3).
32. S. 2(9).
33. *H.C. Deb.*, Standing Committee E., Cols. 26-9 (17 January 1974).
34. Ibid., Col. 30.
35. 314 *H.L. Deb.*, Col. 1079 (28 January 1971); 315 *H.L. Deb.*, Cols. 19, 33, 751 (9 & 18 February 1971); 816 *H.C. Deb.*, Col. 1575 (5 May 1971); 821 *H.C. Deb.*, Col. 109 (12 July 1971).
36. Art. 1(1).
37. The Oil in Navigable Waters (Heavy Diesel Oil) Regulations 1956 (S.I. 1956/897), Reg. 3, replaced by the Oil & Navigable Waters (Heavy Diesel Oil) Regulations 1967 (S.I. 1967/710), Reg. 1.
38. The term 'oil' is defined in s. 1(2) of the Act. The current definition of 'heavy diesel oil' is contained in The Oil in Navigable Waters (Heavy Diesel Oil) Regulations 1967 (S.I. 1967/710) made under the Oil in Navigable Waters Act 1955, as amended in 1963. These Acts are repealed by virtue of s. 33(2). The categories of 'oil' may be extended. See s. 1(2).
39. Part II, ss. 12-16.
40. Art. II(3).

41. See Cusine, 'Liability for Oil Pollution under the Merchant Shipping (Oil Pollution) Act 1971', (1978) 10 J.M.L.C. 105 at 111; and 'The International Oil Pollution Fund as Implemented in the United Kingdom', (1978) 9 J.L.M.C. 495 at 502-3.

42. According to the *Torrey Canyon* (1967) (Cmnd. 3246), the vessel spilled 117,000 tons of crude oil. The report of the proceedings in the United States ([1969] 2 Lloyd's Rep. 591) puts the figure at 119,328 tons. The *Amoco Cadiz* spilled 230,000 tons. Other significant spills are listed below:

Date	Vessel	Location	Spillage (tons)
1970	*Pacific Glory*	Off Isle of Wight	6,200
1974	*Metulla*	Straits of Magellan	59,750
1975	*Olympic Alliance*	Dover Straits	2,000
1976	*Argo Merchant*	New England	25,100
1977	*Venoil/Venpet*	South Africa	25,058
1978	*Eleni V*	Off Suffolk Coast, England	5,000
1978	*Christos Bitas*	Off Pembrokeshire, Wales	3,000

For a fuller list, see IMCO, *News No. 1* (1978), 12-13.

43. Annex C. to the 1971 Amendments.

44. Ss. 10-13 and Sch. 2.

45. S. 11(1). For the purpose of Part II of the Act an 'oil tanker' means 'a ship which is constructed or adapted primarily to carry oil in bulk in its cargo spaces (whether or not it is also constructed or adapted as to be capable of carrying other cargoes in those spaces)'. 'United Kingdom oil tanker' means 'an oil tanker registered in the United Kingdom': s. 10(4).

46. S. 11(1).

47. S. 12(1).

48. Certificates granted by countries which are parties to the Conventions (defined in s. 10(1)) will be recognised by the UK (Sch. 3, para. 3) and certificates issued by non-Convention countries may also be recognised (Sch. 3, para. 4). The countries which are parties to the Conventions will be declared by Order in Council (s. 10(3)) and the current list can be seen in Halsbury's *Statutory Instruments*, Vol. 20, 182-3 and Supplements.

49. Sch. 3.

50. S. 13(1).

51. The Oil in Navigable Waters (Ships' Equipment) (No. 1) Regulations 1956 (S.I. 1956/1423); The Oil in Navigable Waters (Ships' Equipment) Regulations 1957 (S.I. 1957/1424).

52. S. 33(2).

53. Art. VI as added by the 1971 Amendments.

54. S. 11 and Sch. 2.

55. Annex 1, Reg. 9.

56. Annex 1, Reg. 13(1).

57. Protocol to MARPOL, Annex 1., Reg. 13.

58. S. 20, which came into force on 1 August 1979.

59. The Merchant Shipping (Cargo Ship Construction and Survey) (Tanker and Combination Carrier) Rules 1975 (S.I. 1975/750).

60. The Merchant Shipping (Fire Appliances) (Amendment) Rules 1974 (S.I. 1974/2185).

61. In the 1975 Rules, a 'tanker' is defined as 'a cargo ship constructed or adapted for the carriage of crude oil and petroleum products having a closed flash-point not exceeding 60°C. and the Reid vapour of which is below that of atmospheric pressure and other liquids having a similar fire hazard': Rule 1(2). In the 1974 Rules,

which cover 'Category A tankers' and 'Category A combination carriers', i.e. tankers over 100,000 tons deadweight and combination carriers over 50,000 tons deadweight, these terms mean '"a tanker", or as the case may be a combination carrier, registered in the United Kingdom and constructed or adapted to carry crude oil and petroleum products having a closed flashpoint not exceeding 60°C. and the Reid vapour of which is below that of atmospheric pressure and other liquids having a similar fire hazard, and the keel of which: (i) is laid, or which is at a similar stage of construction, on or after 1st February, 1975 or (ii) is laid or is at a similar stage of construction before 1st February 1975 but is completed after 31st December 1978': Rule 3.

62. S. 10(4).

63. IMCO Press Release, 7 March 1968.

64. S.I. 1976/302.

65. Annex 1, Ch. 1. Regs. 12, 29.

66. IMCO Resolution A.321 (1975).

67. Annex 1, Ch. 1. Regs. 6, 19.

68. S. 20, which came into force on 1 August 1979.

69. The Collisions Regulations (Traffic Separation Schemes) Order 1972 (S.I. 1972/809), as amended S.I. 1972/1267 and S.I. 1974/1890.

70. International Regulations for Preventing Collisions, *supra*, Rule 6.

71. Abecassis, *Oil Pollution from Ships*, 59-60.

72. International Conventions for the Safety of Life at Sea 1960 and 1974; Chapter V, Reg. 13.

73. See Abecassis, *Oil Pollution from Ships*, 60-2.

74. The Merchant Shipping (Certificate of Competency as AB) (Regulations) 1970 (S.I. 1970/294).

75. Merchant Shipping Act 1970 ss. 43-51 and subordinate legislation.

76. ACOPS, *Annual Report 1977*, p. 12; *Accidents at Sea causing Oil Pollution: Review of contingency measures* (Dept. of Trade: Marine Division, 1978), p. 13.

77. Convention, Art. 2.

78. Convention, Art. 4(2).

79. Merchant Navy Training Bulletin No. 3-75.

80. E.g. British Petroleum, Shell, Esso.

81. Merchant Shipping Notice No. 771 issued by the Department of Trade Marine Division, July 1976.

82. 'Charterers "Should Ban Sub-Standard Vessels"' (1978) 4 L.M.C.L.Q. 69.

83. ACOPS, *Annual Report 1977*, p. 10. The Panamanian Government recently received assistance from the UK in an attempt to improve the standard of their vessels and crews. See *The Times*, 7 and 9 August 1978.

84. ACOPS, *Annual Report 1977*, p. 10; O.E.C.D. Report, *Flags of Convenience*, IMCO MSC XXVI 16(d); ACOPS, *Annual Report 1978*, 28.

85. ACOPS, *Annual Report 1977*, p. 10; *Annual Report 1978*, p. 28. It was pointed out in the course of the debate on the Merchant Shipping Bill (now the Merchant Shipping Act 1979) that vessels such as *Torrey Canyon*, *Amoco Cadiz*, *Eleni V* and *Christos Bitas*, which had caused extensive pollution had been chartered to major oil companies: *Torrey Canyon* to BP, *Amoco Cadiz* to Shell, *Eleni V* to Chevron and *Christos Bitas* to BP (965 *H.C. Deb.*, Col. 811, 30 March 1979). In the same debate, the Under-Secretary of State for Trade, Mr Stanley Clinton Davis, indicated that oil companies had been asked to monitor their inspection procedures in respect of ships which they propose to charter (965 *H.C. Deb.*, Col. 829, 30 March 1979).

86. The following statement by Mr Walder of the International Oil Companies' Marine Forum was quoted in a parliamentary debate on 'Oil Spillage'. 'The view of our industry which is supported by many independent studies, is that by far the

greatest single cause of tanker accidents is human fallibility. Some 85 per cent. of all navigational accidents can be directly attributed to human failure.' 958 *H.C. Deb.*, Col. 86 (27 November 1978).

87. *The Sunday Times*, 7 May 1978.

88. *The Sunday Times*, 15 July 1978.

89. See Table 1 *supra*.

90. 388 *H.L. Deb.*, Cols. 343-62; 370-401, 'Tanker Safety and Pollution Prevention' (25 January 1978); see especially Lord Ritchie-Calder, Cols. 346-8. Also 395 *H.L. Deb.*, Col. 1009 (27 July 1978), statement by the Earl of Kinnoull.

91. *The Battle Against Oil Pollution at Sea* (Marine Division of Department of Trade, 1976), 3.

92. Art. III(b).

93. Reg. 1 and Appendix 1.

94. Reg. 9(1)(a)(v).

95. Reg. 9(1)(a)(i) and Reg. 10.

96. Reg. 9(1)(a)(vi). See also Reg. 15.

97. Reg. 12.

98. S.I. 1972/1928.

99. *Federal Steam Navigation Co. Ltd. & Anr. v. Dept. of Trade and Industry* [1974] 2 All E.R. 97.

100. S.2(4).

101. S. 5.

102. S. 5(2).

103. See the decision of Sheriff Isobel Poole in *P.F. v. Mobil North Sea Ltd*, Aberdeen Sheriff Court: *The Scotsman*, 17 July 1979, 7.

104. Art. IX and Annex.

105. Reg. 20 and Annex III.

106. The Oil in Navigable Waters (Records) Regulations 1972 (S.I. 1972/1929).

107. The Oil in Navigable Waters (Transfer Records) Regulations 1957 (S.I. 1957/358), continued in force by virtue of s. 33(2) of the Prevention of Oil Pollution Act.

108. S. 18.

109. S. 21.

110. S. 19.

111. S. 20.

112. ACOPS, *Annual Report 1977*, 29: *Annual Report 1978*, 59.

113. Ibid., 32.

114. See Table 1 *supra*.

115. *Uncontrolled Blow-out on Bravo*, Report from the Commission of Inquiry appointed by the Royal Decree of 26 April 1977, 28. Discussed more fully by Fleischer in Ch. 5 of this book.

116. Art. 24.

117. Art. 5(7).

118. Young, 'Legal Status of Submarine Areas Beneath the High Seas' (1951) 45 A.J.I.L. 225.

119. Discussed *supra*.

120. Oil Development Council for Scotland, *North Sea Oil and the Environment*, 3-6.

121. Continental Shelf Operations Notice No. 37, Dept. of Energy August, 1977; *Oil Spill Clean Up Arrangements in the Offshore Oilfields* (Depts. of Trade and Energy, March 1978). These plans are available for inspection. See ACOPS *Annual Report 1978*, 19.

122. S. 1.

123. S. 2.

124. The Continental Shelf (Designation of Areas) Order 1964, S.I. 1964/697; The Continental Shelf (Designation of Add. Areas) Orders, S.I. 1965/1531, 1968/ 891, 1971/594, 1974/1489, 1976/1153, 1977/1871, 1978/178.

125. The current regulations are the Petroleum (Production) Regulations 1976, S.I. 1976/1129, consolidating, with amendments, earlier regulations: S.I. 1966/ 898, 1971/814, 1976/276. The 1976 regulations have themselves been amended: S.I. 1978/929.

126. S. 34.

127. Continental Shelf Act 1964, s. 2; The Continental Shelf (Protection of Installations) Orders 1976-7; S.I. 1976/332, 954, 1308, 1497; S.I. 1977/712, 966, 1035, 1344.

128. 885 *H.C. Deb.*, Cols. 220-1; 222-3 (11 February 1975).

129. Landward Production Model Clauses (L.P.M.C.), clause 21(1); Seaward Production, M.C. (S.P.M.C.), cl. 21(1); Exploration M.C. (E.M.C.), cl. 10.

130. Part VIII entitled 'Code of Safe Practice for Drilling and Production in. Marine Areas'.

131. L.P.M.C., cl. 21(2); S.P.M.C., cl. 21(2); E.M.C., cl. 10(2).

132. L.P.M.C., cl. 21(1) (e); S.P.M.C., cl. 21(1) (e); E.M.C., cl. 10(1) (a).

133. L.P.M.C., cl. 21(8); S.P.M.C., cl. 21(8); E.M.C., cl. 10(3).

134. L.P.M.C., cl. 23; S.P.M.C., cl. 23; E.M.C., cl. 11.

135. L.P.M.C., cl. 17(4) & (5); S.P.M.C., cl. 17(4) & (5); E.M.C., cl. 8.

136. L.P.M.C., cl. 20; S.P.M.C., cl. 20.

137. For a fuller discussion of the role of insurance, see Mankabady, Ch. 6 of this book.

138. The text of the original Agreement is to be found in (1974) 13 I.L.M. 1409 and the Rules in (1975) 14 I.L.M. 147. See further, Ch. 6.

139. L.P.M.C., cl. 38; S.P.M.C., cl. 40; E.M.C., cl. 22.

140. *Accidental Oil Pollution of the Sea* (Department of the Environment) (Central Unit on Environmental Pollution Paper No. 8, HMSO, 1976), para. 6.19.

141. Ss. 2, 3, 4, 6, 7.

142. The Offshore Installations [henceforth OI] (Registration) Regulations 1972, S.I. 1972/702; OI (Managers') Regulations 1972, S.I. 1972/703; OI (Construction & Survey) Regulations 1974, S.I. 1974/289; OI (Logbooks and Registration of Death) Regulations 1972, S.I. 1972/1542; OI (Inspectors & Casualties) Regulations 1973, S.I. 1973/1842; OI (Public Inquiries) Regulations 1974, S.I. 1974/338; OI (Diving Operations) Regulations 1974, S.I. 1974/1229; OI (Application of the Employers' Liability (Compulsory Insurance) Act 1969) Regulations 1975, S.I. 1975/1289; OI (Emergency Procedures) Regulations 1976, S.I. 1976/1542; OI (Operational Safety, Health and Welfare) Regulations 1976, S.I. 1976/1019; OI (Life-saving Appliances) Regulations 1977, S.I. 1977/486; OI (Fire-fighting Equipment) Regulations 1978, S.I. 1978/611.

143. The Submarine Pipelines (Diving Operation) Regulation 1976, S.I. 1976/ 923; The Submarine Pipelines (Inspections) Regulations 1977, S.I. 1977/835.

144. We are indebted to Mr A.D. Read of the Petroleum Engineering Division of the Dept. of Energy for this information.

145. The Offshore Installations (Registration) Regulations 1972.

146. The Offshore Installations (Construction and Survey) Regulations 1974, Reg. 3.

147. Ibid., Reg. 7.

148. Ibid., Reg. 9(2).

149. Ibid., Reg. 11(1).

150. Ibid., Reg. 8(2) (a).

151. The Offshore Installations (Managers) Regulations 1972.

152. The Offshore Installations (Logbooks and Registration of Death) Regulations 1972, Reg. 3(2) (d).

153. The Offshore Installations (Inspectors and Casualties) Regulations 1973, Regs. 2, 4-7.

154. Ibid., Reg. 2(1) (8).

155. Ibid., Reg. 7.

156. See Table 1.

157. Brown, 'The Lessons of the Torrey Canyon: International Law Aspects', (1968) 21 Cur. Leg. Prob. 113.

158. The 1973 Protocol to the Intervention Convention extends its provisions to substances which may be designated by IMCO and, more generally, to 'those other substances which are liable to create hazards to human health, to harm living resources and marine life, to damage amenities or to interfere with other legitimate uses of the sea'.

159. Art. 1.

160. Art. 3.

161. Art. 5.

162. Ss. 12, 16.

163. S. 12(1).

164. S. 12(2); (3).

165. S. 12(4).

166. S. 14.

167. S. 13.

168. S. 12.

169. Art. I(1).

170. Art. II(1).

171. Art. II.

172. *Salomon* v. *Commissioners of Customs & Excise* [1967] 2 Q.B. 116; see *Maxwell on Interpretation of Statutes* (12th edn), 183-6.

173. CUEP, *Accidental Oil Pollution of the Sea*, Pollution Paper No. 8 (1976), para. 11. 51.

174. S. 1(1).

175. S. 5(5).

176. S. 5(7).

177. S. 5(2).

178. S. 29(1).

179. Art. II(3).

180. Art. I.

181. Local Government Act 1972, s. 138; Local Government (Scotland) Act 1973, s. 84; Coast Protection Act 1949, s. 4.

182. *Accidental Oil Pollution of the Sea, supra*, para. 11. 47.

183. *Coast Pollution: Observations on the Report of the Select Committee on Science and Technology* (1969, Cmnd. 3880), para. 30.

184. Circular 34/68.

185. Circular 29/68.

186. Circular 55/68.

187. *Accidental Oil Pollution of the Sea, supra*, para. 11. 50.

188. Ibid., para. 52: see also ACOPS, *Annual Report 1978*, 8-9.

189. *Accidents at Sea causing Oil Pollution. Review of Contingency Measures* (Department of Trade, Marine Division, 1978), Report of a Steering Committee.

190. Ibid., para. 57.

191. Ibid., para. 56.

192. *Review of Contingency Measures*, para. 47.

193. 958 *H.C. Deb.*, Written Answers, Cols. 474-475 (20 November 1978).

194. S.I. 1976/1542.

195. *Glasgow Herald*, 23 August 1978.

196. We are indebted to Mr A.D. Read of the Petroleum Engineering Division of the Dept. of Energy for this information.

197. See generally *Torrey Canyon* (1967, Cmnd. 3246).

198. [1956] A.C. 218, discussed *supra.*

199. The 1894 Act, s. 503, as amended by the 1958 Act, s. 1.

200. Keaton, 'The Lessons of the "Torrey Canyon": English Law Aspects', (1968) 21 Cur. Leg. Prob. 94.

201. S. 1.

202. S. 4.

203. Art. IX(2).

204. S. 13.

205. S. 14(3).

206. S. 13(2).

207. S. 9.

208. Art. III(1).

209. Defined in Art. 1(3) as 'the person or persons registered as the owner of the ship or, in the absence of registration, the person or persons owning the ship. However, in the case of a ship owned by a State and operated by a company which in that State is registered as the ship's operator "owner" shall include such company'.

210. Defined in Art. I(1) as 'any sea-going vessel or any seaborne craft of any type whatsoever actually carrying oil in bulk as cargo'. The provisions of the Convention do not apply to warships or other ships owned or operated by a State and used, for the time being, on government non-commercial service: Art. XI.

211. Defined in Art. I(6) as 'loss or damage caused outside the ship carrying oil by contamination resulting from the escape or discharge of oil from the ship, wherever such escape or discharge may occur, and includes the costs of preventive measures and further loss or damage caused by preventive measures:' '"Preventive measures" means any reasonable measure taken by any person after an incident has occurred to prevent or minimise pollution damage': Art. I(7).

212. Defined in Art. I(5) as 'any persistent oil such as crude oil, fuel oil, heavy diesel oil, lubricating oil and whale oil, whether carried on board a ship as cargo or in the bunkers of such a ship'.

213. Defined in Art. I(8) as 'any occurrence, or series of occurrences having the same origin, which causes pollution damage'.

214. Art. II.

215. The definition of 'person' in Art. I(2). Liability is dealt with in Arts. III and IV. In the corresponding section of the Act (s. 1) the owner is made liable and there is no mention of 'the person' to whom he is liable. That being so, anyone alleging that he has suffered damage must, on existing common law principles, show that he has both title and interest to sue.

216. Art. IV.

217. Art. III(2) (a).

218. Art. III(2) (b).

219. Art. III(3).

220. Art. III(2) (c).

221. Art. VIII(1).

222. Art. VII(2). The details of the certificate are dealt with in Art. VII(2) and (3).

223. Art. VII(4).

224. S. 1.

225. The term is defined in s. 20(1): '"owner" in relation to a registered ship, means the person registered as its owner, except in relation to a ship owned by a State which is operated by a person registered as its operator'. The definition

in the CLC is slightly wider in that it includes the owner of a ship which is not registered (Art. I(3)). This is probably not significant in the UK in that, with few exceptions, all ships above 15 tons must be registered (the 1894 Act, ss. 2-3). Both definitions exclude the bareboat charterer or charterer by demise. For the purposes of day-to-day management of the ship, however, he is the equivalent of the owner: Carver, *Carriage by Sea* (12th edn), paras. 318-23.

226. Since the 1971 Act is to be construed as one with the Merchant Shipping Acts 1894-1970 (s. 21(2)), a '"ship" includes every description of vessel used in navigation not propelled by oars' (1894 Act, s. 742 as amended). The 1971 Act also applies to hovercraft (s. 17), but it excludes warships or other ships used by the government of a State for non-commercial purposes (s. 14(1)). Other ships are excluded by s. 8A which was inserted in the 1971 Act by s. 9 of the 1974 Act. Because of the above definition, the Act, unlike the CLC (Art. I(1)), is not restricted to 'sea-going vessels'.

227. The ship must be carrying a cargo of persistent oil, but the oil which is discharged or escapes need not come from the cargo. The Act does not apply to the escape or discharge of oil on a ballast run.

228. Not defined in the Act. The CLC uses the word 'incident' (Art. I(8)) which it defines as 'any occurrence, or series of occurrences having the same origin, which causes pollution damage'. The Act reflects this in s. 1(4).

229. '"Damage" includes loss' (s. 20(3)). There is a fuller definition in the CLC (Art. I(6), (7)). Although the Act talks about 'any damage', some limit would be placed on losses for which compensation would be available by the operation of the common law notion of 'remoteness'. See *Salmond on Torts* (16th edn), 551-5.

230. Reference to the 'area' of a country includes its territorial waters (s. 20(3)).

231. The countries which are parties to the CLC will be declared by Order in Council (s. 19(2)). The present Orders are the Merchant Shipping (Oil Pollution) (Parties to Convention) Order 1975, S.I. 1036, 1975; the Merchant Shipping (Oil Pollution) (Parties to Convention) (Amendment No. 2) Order 1977, S.I. 1977/826.

232. Under the 1974 Act, s. 5, the owner may recover part or all of the cost of his preventive or 'clean-up' measures.

233. S. 1(3).

234. S. 3.

235. Art. III(2) (b).

236. 821 *H.C. Deb*, Cols. 109-12 (12 July 1971).

237. S. 3. exempts them only from liability under s. 1.

238. Ss. 4-5.

239. Limitation was originally available only to the owner. The 1958 Act extended it to 'any charterer and any person interested in or in possession of the ship, and, in particular any manager or operator of the ship': s. 3(1).

240. Forster, 'Civil Liability of Shipowners for Oil Pollution', 1973 J.B.L. 23.

241. S. 2.

242. Art. III(2).

243. See Forster, 'Civil Liability' and Cusine, 'Liability for Oil Pollution under the Merchant Shipping (Oil Pollution) Act 1971', *supra*.

244. S. 1(5).

245. The CLC will relieve the owner from liability only if the claimant intended to cause damage, which is not the same as contributory negligence.

246. Art. V(1).

247. The term 'franc' is defined in Art. V(9).

248. Defined in Art. V(10) as 'the net tonnage of the ship with the addition of the amount deducted from the gross tonnage on account of engine room space for the purpose of ascertaining the net tonnage. In the case of a ship which cannot

be measured in accordance with the normal rules for tonnage measurements, the ship's tonnage shall be deemed to be 40% of the weight in tons (of 2240 lbs.) of oil which the ship is capable of carrying'. Under section 503(2)(a) of the 1894 Act and section 69 of the 1906 Act, the shipowner may limit his liability on the basis of the ship's tonnage, the definition of which is the same as that in the Convention. Regulations on ship's tonnage are made under s. 1 of the 1965 Act.

249. See, for example, Temperley, *The Merchant Shipping Acts* (7th edn), para. 431. When the Convention on Limitation of Liability for Maritime Claims 1976 ((1977), 16 I.L.M. 606) comes into force, the term 'actual fault or privity' will be replaced by 'personal act or omission'. See ss. 17 and 18 of the 1979 Act.

250. Art. V(2).

251. Art. V(3).

252. Art. V(11).

253. Art. VII(8).

254. Art. V(3).

255. S. 4.

256. This phrase and its problems are familiar to maritime lawyers. See Temperley, *The Merchant Shipping Acts*, para. 431.

257. S. 5(1).

258. The term 'gold franc' is defined in s. 4(3); See Mendelsohn, 'The Value of the Poincaré Gold Franc in Limitation of Liability Conventions', (1974) J.M.L.C. 125. The gold franc has now been replaced by Special Drawing Rights on the International Monetary Fund: see the Protocols to the CLC and the Fund Convention (1977) 16 I.L.M. 617 and 621; and Bristow, 'Gold Franc-Replacement of Unit of Account', [1978] 1 L.M.C.L.Q. 31.

259. S. 4(2) deals with the calculation of tonnage.

260. S. 4(1)(b).

261. S. 4(4). The current equivalents are set out in the Merchant Shipping (Sterling Equivalents) (Various Enactments) Order 1979, S.I. 1979/790.

1971 Act

2,000 gold francs is now equal to	£81.97
210m	£8,606,346.00

1974 Act

1,500 gold francs is now equal to	£61.47
2,000	£81.97
125m	£5,122,825.00
250m	£8,606,346.00
450m	£18,442,170.00
675m	£27,663,255.00
900m	£36,884,340.00

Notes (1) These sterling equivalents are calculated with reference to the Special Drawing Rights on the I.M.F.

(2) On 27 April 1979, the Assembly substituted the figure of 675m gold francs for that of 450m in Art. 4(4) of the Fund Convention (see Art. 4(6)).

262. S. 8A inserted in the 1971 Act by s. 9 of the 1974 Act. This will be repealed when the relevant provisions of the 1979 Act come into force (s. 50(4) and Sch. 7).

263. S. 5(2).

264. S. 5(2)(b); (3)-(4); The method of apportionment is the same as that under s. 503 of the 1894 Act as amended. For details, see Temperley, *The Merchant*

Shipping Acts, paras. 437-46.

265. S. 4(5).
266. S. 6(1)(a).
267. S. 6(1)(b).
268. J. Graham Stewart, *A Treatise on the Law of Diligence* (W. Green & Sons, Edinburgh, 1898), 88 *et seq.*
269. *Rules of Court*, 1965, Order 22, Rule 3(4), (5).
270. This is expressly stated in the CLC, Art. VI(1)(a).
271. By virtue of s. 12.
272. S. (4).
273. S. 5(5). This is a set-off.
274. Art. X(1)(b).
275. S. 13(3).
276. Art. VII.
277. S. 10.
278. S. 11(1).
279. S. 11(2).
280. S. 11(3)(b). Reg. 6(2) of the above Regulations.
281. United States Senate Foreign Relations Committee, Subcommittee on Oceans and International Environment, 'Conventions and Amendments Relating to Pollution of the Seas by Oil', *92nd Congress*, 1st Session, 222-4.
282. S. 10(2).
283. S. 10(8).
284. The use of the conjunction 'and' makes it clear that both may be convicted. In a similar section, the Prevention of Oil Pollution Act 1971 uses the word 'or' (s. 1(1), 2(1)). However, when the House of Lords were asked to construe the word 'or' in the Oil in Navigable Waters Act 1955 s. 1(1) (which is repealed and replaced by the 1971 Act), they decided that the word was used conjunctively and not disjunctively. *Federal Steam Navigation Co. Ltd. and Anr.* v. *Dept. of Trade and Industry* [1974] 2 All E.R. 97. The owner and master may avail themselves of the defences in the Act. The defences to a charge under s. 12 are contained in s. 12(5) and the defences to a charge under s. 13 are in s. 13(5).
285. S. 10(6).
286. S. 10(7). As from 1 January 1980, this will be increased to £500 in terms of s. 43 of the 1979 Act.
287. S. 12(1). This scheme is similar to that under the Third Parties (Rights against Insurers) Act 1930, but its provisions are specifically excluded by the 1971 Act (s. 12(5)).
288. S. 12(3).
289. S. 12(2).
290. Civil Evidence Act 1968 s. 11; Law Reform (Miscellaneous Provisions) (Scotland) Act 1968, s. 10.
291. *P. Samuel & Co. Ltd.* v. *Dumas* [1924] A.C. 431.
292. S. 4(1).
293. Art. X.
294. Annotations to the Merchant Shipping Act 1974 by Charles MacDonald (*Current Law Statutes 1974*).
295. S. 2(8).
296. S. 3.
297. The circumstances in which this will occur are set out in s. 4(1) of the 1974 Act.
298. S. 4(7)(b).
299. ACOPS, *Annual Report 1976*, 8-9; *Annual Report 1977*, 7-8; *Annual Report 1978*, 29.

300. S. 4(7) (a) (ii).
301. S. 4(7) (a) (i).
302. Forster, 'Civil Liability', *supra,* 25-26.
303. S. 2.
304. S. 5(1). The Fund will relieve them of that part of their liability which is in excess of an amount equivalent to the lesser of 1,500 francs per ton or 125 million francs, but not in excess of an amount equivalent to the lesser of 2,000 francs per ton or 210 million francs.
305. S. 5(3).
306. S. 5(4).
307. The Fund Convention (Art. 5(3)(a)) requires compliance with certain International Conventions. These are the 1954 Convention or SOLAS 1960 or the International Convention of Load Lines 1966 or the International Regulations for Preventing Collisions at Sea and any important amendments thereof. The requirements under the 1974 Act are to be found in the Merchant Shipping (Indemnification of Shipowners) Order 1978 (S.I. 1978/1467).
308. S. 7(1).
309. S. 7(2).
310. S. 7(3).
311. S. 13.
312. S. 6(1).
313. S. 6(4)(5).
314. Discussed more fully by Becker, Ch. 4 of this book.
315. Anton, *Private International Law,* 245-8.
316. Graveson, *Conflict of Laws* (7th edn), 574-7.
317. S. 11(2).
318. S. 11(7).
319. Discussed by Fleischer, Ch. 5 and Mankabady, Ch. 6 of this book.
320. See Table 1.
321. See, e.g. Cusine, 'Liability for Oil Pollution under the Merchant Shipping (Oil Pollution) Act 1971'. (1978) 10 J.M.L.C. 105; and 'The International Oil Pollution Fund as Implemented in the United Kingdom' (1978) 9 J.M.L.C. 495.
322. ACOPS, *Annual Report 1976,* 8-9; *Annual Report 1977,* 7-8; *Annual Report 1978,* 29.
323. ACOPS, *Annual Report 1977,* 14. In 1978, out of 507 incidents, 242 were 'unattributed'. See ACOPS, *Annual Report 1978,* Table 4, 49. The Group on Liability and Compensation has recommended that priority be given to quantifying the costs incurred in connection with such spills so that the UK Government may consider what further action should be taken. 'Liability and Compensation for Marine Oil Damage', para. 40.
324. *New York Times,* 8 November 1975, 1. 'Fingerprinting' is a chemical analysis of the oils to identify it with a particular tanker and 'tagging' involves introducing into the oil a trace substance so that a tanker's oil can be identified. These are not foolproof, as some tankers may take on several different types of oil from various ports. The Marine Environment Protection Committee (a committee of IMCO) is to investigate means of identifying the sources of discharged oil. See *Lloyd's Shipping Economist,* Vol. 1, No. 7, 14.
325. See Birnie, Ch. 3 of this book.
326. IMCO Briefing dated 7 November 1979 (IMCO/B24/79). We are grateful to Roger Kohn the Information Officer for IMCO for these details.
327. See Abecassis, *Oil Pollution from Ships,* 18-19.
328. Fourth Report from the Select Committee on Science and Technology, 'Eleni V' (1978) *H.C. Paper* 684.
329. See ACOPS, *Annual Report 1978,* 11.

2 THE ROLE OF IMCO IN COMBATING MARINE POLLUTION

R.R. Churchill

The Inter-Governmental Maritime Consultative Organisation (IMCO) is the United Nations specialised agency dealing with shipping matters. As an organisation, IMCO has had a rather chequered history. It was set up by a Convention which was adopted and opened for signature at a UN maritime conference held in Geneva in March 1948. It was 10 years, however, before the Convention received the 21 ratifications necessary for it to come into force and for IMCO to be formally established. There appear to be several reasons for this delay: doubts that an inter-governmental agency could achieve a great deal in the shipping industry, dominated as it is by private interests; fears by a number of important shipping States that IMCO would involve itself in commercial matters; and fears by States with smaller shipping industries that the institutional structure of IMCO would prove undemocratic. Nevertheless, once the 21 ratifications had been deposited, those States with shipping industries of any consequence which had not become members were quick to join. One group of States which were in no hurry to become members were those Third World States which became independent in the 1960s. They regarded IMCO as an organisation existing largely for the benefit of industrialised shipping countries and having little to offer them. Their main concern was to build up shipping industries of their own and to gain access to the lucrative liner conferences which divide up trade in cargoes between different shipping lines. For these purposes the UN Conference on Trade and Development (UNCTAD) was regarded as a much more useful organisation than IMCO. Nevertheless, by the early 1970s, these Third World States began to realise that IMCO did have something to offer them in the form of technical assistance and in other ways. Since 1973 no less than 30 Third World States have become members. Today IMCO has 110 member States and one associate member (Hong Kong), compared with a total UN membership of 151. Those UN members which are not members of IMCO are mainly some land-locked and smaller Third World States. Thus, from its rather in-auspicious beginnings, through its cautious acceptance by shipping nations, IMCO has arrived at a position where it has received world-wide acceptance, thanks in large part to its solid technical achievements. It is because of its early history that IMCO has consistently sought to maintain a fairly low profile and to concentrate almost exclusively on technical matters.

The initial doubts and reservations about setting up IMCO can be seen in the modesty of the purposes and functions conferred on the organisation. Art. 1 of IMCO's constituent Convention provides that its main purposes are to provide machinery for inter-governmental co-operation in relation to the technical regulation of shipping, in particular to encourage the general adoption of the highest practicable standards in maritime safety and efficiency of navigation, and to consider any matter concerning shipping that may be referred to it by any UN body. Art. 1 also includes in the list of IMCO's purposes encouragement of the removal of discriminatory practices in inter-national shipping and consideration of unfair restrictive practices by shipping concerns, but in practice these purposes have been virtually ignored. In relation to its purposes, IMCO may make recommendations, provide for the drafting of Conventions and recommend them to governments, and provide machinery for consultation and exchange of information among members.

Thus, it can be seen that there is no mention of marine pollution anywhere among IMCO's purposes or functions. This is not surprising, since at the time of the drafting of the IMCO Convention in 1948, there was little concern with marine pollution. In practice, however, the lack of any mention of marine pollution has not prevented IMCO from con-sidering and taking a considerable amount of action on the question: IMCO's purposes are sufficiently broadly formulated – particularly the reference to 'any matters concerning shipping' – to cover marine pollution. It is interesting to note that Amendments adopted to the IMCO Con-vention in 1975 (which are expected to come into force sometime in the early 1980s) amend the list of purposes of the organisation so that they now include the encouragement of 'the general adoption of the highest practicable standards in matters concerning . . . the prevention and control of marine pollution from ships; and to deal with legal matters related' thereto. In the early days of its existence marine pollution was not one of IMCO's major concerns. It was only after the *Torrey Canyon* disaster[1] in March 1967 that IMCO became really involved in the question of controlling and preventing marine pollution. In the dozen years that have elapsed since the *Torrey Canyon* IMCO has accomplished a good deal in this field. It is IMCO's work – particularly its legal aspects – concerning the prevention and control of marine pollution from ships (and this is the only source of marine pollution with which IMCO is concerned) which this chapter attempts to review. IMCO's work in this field (which is of course only part of its overall work programme) will be dealt with under three main headings: (i) conventions dealing directly with marine pollution; (ii) conventions which may indirectly help to prevent marine pollution; and (iii) other measures to combat marine pollution. First, however, it is necessary to say a few words about the institutional structure of IMCO.

1 INSTITUTIONAL STRUCTURE OF IMCO

In keeping with IMCO's modest beginnings and its rather pragmatic existence thereafter, its institutional structure was initially very simple, with a number of new organs being added only when this was considered necessary. Today the main organs are the Assembly, Council, Maritime Safety Committee, Legal Committee, Marine Environment Protection Committee and the Secretariat. The Assembly consists of all member States and normally meets once every two years. It is the supreme governing body of the Organisation. Its main functions are to approve the Organisation's work programme and budget, to make non-binding recommendations to member States relating to maritime safety and, under the 1975 Amendments referred to above, relating to the prevention and control of marine pollution from ships, and (again under the 1975 Amendments) to take decisions in regard to convening any international conference or following any other appropriate procedure for the adoption of international conventions developed by IMCO's committees. The main tasks of the Council, which consists of 24 member States elected by the Assembly and which meets as often as necessary, are, subject to the authority of the Assembly, to supervise the execution of IMCO's work programme and between sessions of the Assembly to perform all the functions of the Organisation, except the Assembly's function of making recommendations (referred to above). The Maritime Safety Committee consists of all IMCO members and meets two or three times a year. The Committee considers questions of maritime safety and navigation and, through the Council, makes proposals to the Assembly for safety regulations. The Legal Committee was set up in 1967 following the *Torrey Canyon* disaster. It consists of all member States and meets at least once a year. Its main tasks are to consider any legal matter within the scope of the Organisation and to draw up draft conventions. The specialist body dealing with marine pollution is the Marine Environment Protection Committee (MEPC). This Committee was not established until November 1973: prior to this, marine pollution had been dealt with by the Maritime Safety Committee. The MEPC executes and co-ordinates all IMCO activities relating to marine pollution and liaises with other international organisations dealing with marine pollution. In particular, it performs such functions as are conferred on IMCO by any Convention on marine pollution; considers appropriate measures to facilitate the enforcement of such conventions; provides for the acquisition and dissemination of scientific and technical information on marine pollution to States, particularly developing countries; promotes co-operation with regional organisations concerned with marine pollution; and (under the 1975 Amendments to the IMCO Convention) submits to the Council proposals for regulations for the prevention and control of marine pollution from ships. Finally there is the Secretariat, consisting of the Secretary-General and a staff of just over 200 – a very small figure when compared with most other UN specialised agencies.

It can be gathered from this that IMCO has no law-making powers. Recommendations made by the Assembly on matters of maritime safety and marine pollution are not binding. IMCO's role is limited to drafting international Conventions and to organising diplomatic confer-ences for their adoption. Such Conventions only enter into force and become binding when they have been signed and ratified by member States of IMCO: and in this matter the member States have an absolute discretion as to whether they ratify a particular Convention. IMCO can-not force any member State to become a party to any Convention, although, of course, it can – and in fact does – urge States which have not yet become parties to international Conventions concerned with the prevention of marine pollution to do so.

The question of Amendments to Conventions is, however, somewhat different. Before 1972 the typical procedure was for the IMCO Assembly to adopt Amendments to Conventions which had been drafted by IMCO and adopted under its auspices. But such Amendments did not become effective and binding until they had been accepted by two-thirds of the States parties to the Convention concerned: once, however, they had been so accepted, they came into force and bound all the parties to the Convention and not merely those parties which had accepted the Amend-ments. The experience of this procedure – and this was true of all IMCO Conventions and not just those concerned with marine pollution – was not particularly happy, as there were often lengthy delays (often upwards of five years) before the Amendments to a particular Convention entered into force. Therefore, since 1972, IMCO has introduced a new procedure which, it is hoped, will lead to Amendments coming into force more rapidly after they have been adopted. The typical feature of this new procedure is to make Amendments to substantive provisions of a

Convention subject to the procedure outlined above, but to provide that Amendments to the technical regulations contained in a Convention (which in practice will be that part of the Convention requiring relatively frequent and rapid adjustment so as to keep up with technological developments) can be made by the IMCO Assembly and will be binding on all States parties to the Convention unless they specifically object. (This is known as the tacit Amendment procedure). In relation to this latter kind of Amendment the IMCO Assembly can be said to have a quasi-legislative role.

Having looked at the institutional structure of IMCO and its general absence of any law-making powers, we must now turn to look at the Conventions dealing with marine pollution from ships which have either been adopted under the auspices of IMCO or in which IMCO plays some other role.

2 CONVENTIONS DEALING DIRECTLY WITH MARINE POLLUTION

In this section we will look at those Conventions dealing directly with marine pollution from ships which have been drafted by IMCO, adopted under its auspices or in which the organisation plays some other role. The aim here will be not so much to give an account of the substantive provisions of these Conventions (which has already been done in Chapter 1), but to concentrate on the part that IMCO plays in them. From the point of view of their subject-matter, the Conventions concerned with marine pollution from ships can be divided into three broad categories: (i) those Conventions concerned with deliberate pollution from ships; (ii) Conventions concerned with liability; and (iii) Conventions concerned with intervention by coastal States against ships causing pollution off their coasts.

2.1 Conventions Concerned with Deliberate Pollution

2.1.1 The 1954 Convention and its Amendments

The International Convention for the Prevention of Pollution of the Sea by Oil was the first international Convention to deal with marine pollution. The aim of the Convention is to reduce deliberate pollution of the sea by oil from ships (which results, for example, from the practice of washing out at sea the empty tanks of oil tankers), by setting maximum discharge standards and by prohibiting the discharge of oil in certain areas altogether. The Convention was adopted at a conference organised by the UK Government in May 1954: as we have already seen, IMCO had not come into existence at this time. Nevertheless the Convention provided for certain functions to be undertaken by IMCO when it came into existence. In fact the IMCO Convention entered into force a few months before the 1954 Convention did, so that IMCO was able to play its role under the Convention from the start. The Convention gives IMCO three main tasks: first, to act as depositary, i.e. to receive ratifications to the Convention. Second, it is to be notified by each State party as to which of its ports have reception facilities for oily waste: this information can then be circulated to the other contracting parties. Third, IMCO is to be sent by each State party the text of its legislation giving effect to the Convention and reports of the way in which it has enforced the Convention. There are some indications that this procedure has not realised its full potential. Many States have not sent in the required legislation or reports. Had they done so, it would have been possible for IMCO to have spotted at least some of the undoubted failures there have been to implement and enforce the Convention properly, and then to have applied political and diplomatic pressure on the recalcitrant States concerned to carry out their obligations under the Convention effectively.

The 1954 Convention has been ratified by 59 States. This may not seem a particularly large figure (given that there are over 150 States), but these 59 States collectively own over 90 per cent of the total world tonnage of oil tankers (at which the Convention is principally aimed). Nevertheless, the Convention has not been very successful in preventing deliberate oil pollution. This is mainly because it has not been adequately enforced, partly as a result of the difficulty in detecting offences and partly as a result of some States (particularly flags of convenience States[2]) being unwilling or unable to take proper action against their offending vessels.

The Convention has been amended three times – in 1962, 1969 and 1971. The 1962 Amendments tightened up discharge standards and increased the number of prohibited areas. The 1969 Amendments have the effect of prohibiting oil discharge through normal operation of a ship except under certain very strict conditions. The object of the 1971 Amendments is to limit oil pollution resulting from the stranding or collision of an oil tanker by setting compulsory limits on the size of tankers (thus for the first time introducing the use of construction standards as an instrument of pollution control). The 1962 Amendments were adopted at a conference organised by IMCO, while the 1969 and 1971 Amendments were drafted by the Maritime Safety Committee and adopted by the IMCO Assembly. In all three cases the Amendments required ratification by two-thirds of the States parties to the 1954

Convention before they entered into force. It was pointed out above
that one of the drawbacks with this form of amendment procedure is
that it often leads to lengthy delays in Amendments coming into force.
The Amendments to the 1954 Convention have been no exception. The
1962 Amendments took just over five years to come into force, the
1969 Amendments over eight years, while the 1971 Amendments have
not yet come into force (they have so far received 23 of the 40 ratifications
necessary, although these ratifications do represent about 65 per cent of
the total world tonnage of oil tankers).

2.1.2 The 1972 London Dumping Convention

The Convention on the Prevention of Marine Pollution by Dumping of
Waste and Other Matter provides a comprehensive legal framework to
control the dumping of industrial and other wastes at sea by ships and
aircraft. The Convention prohibits the dumping of some of the more
noxious wastes and makes the dumping of other wastes subject to a
permit. IMCO, in fact, had nothing to do with the adoption or drafting
of this Convention, which was a product of preparations for the UN
Conference on the Human Environment held in Stockholm in June 1972.
IMCO became involved, however, when after the Convention entered
into force in August 1975 the contracting parties met to designate a
competent organisation to be responsible for secretariat duties in relation
to the Convention (as they were required to do) and chose IMCO for
this purpose. IMCO's secretariat duties under the Convention include:
(a) convening consultative meetings of the contracting parties not less
frequently than once every two years; (b) helping to develop criteria
and procedures for the dumping of otherwise prohibited waste in
emergencies; (c) consulting with the contracting parties and making
recommendations to them on questions related to, but not specifically
covered by, the Convention; (d) circulating to contracting parties
certain information which each party must notify to it, for example
records of the waste it permits to be dumped; (e) collaborating with
contracting parties in giving support to those parties which require help
in the training of personnel, the supply of equipment for research and
monitoring, and in the disposal and treatment of waste; (f) acting as
the depositary for ratifications made to any Amendments adopted to
the Convention. There is at present little information on the way in which
IMCO is carrying out these duties in practice.

2.1.3 The 1973 Convention on Pollution from Ships (MARPOL) with its Protocol of 1978

The aim of MARPOL is to control all forms of international pollution of the sea from ships other than the dumping of wastes. It was drafted by IMCO and was adopted at a diplomatic conference held under IMCO's auspices in London in October-November 1973. IMCO also acts as the depositary for ratifications to the Convention. If and when MARPOL comes into force, there are a number of functions for IMCO to perform. These include: (a) being notified by contracting parties of the way in which they are implementing and enforcing the Convention, of their reception facilities and their agencies dealing with ship design and construction, of any incidents involving harmful substances or casualties to their ships causing serious pollution, and of any dispute submitted to arbitration in accordance with the terms of the Convention; (b) to help in the promotion of technical co-operation relating to pollution control; and (c) to adopt Amendments to the Convention when required: in the case of Amendments relating to the technical regulations of the Convention, the tacit amendment procedure, outlined above, applies.

There is no doubt that were the Convention to come into force and be widely ratified, it would make a significant contribution to reducing pollution from ships. Unfortunately, however, it is making very slow progress at coming into force. To do so, it requires ratification by 15 States, the combined merchant fleets of which constitute not less than 50 per cent of the gross tonnage of the world's merchant shipping. Five years after the adoption of the Convention, at the end of 1978, it had received only three ratifications—those of Jordan, Kenya and Tunisia, the combined merchant fleets of which amount to only 0.03 per cent of the total world merchant fleet!

Partly in an effort to try and speed up the entry into force of MARPOL, IMCO organised a Conference on Tanker Safety and Pollution Prevention which was held in London in February 1978 and which adopted a Protocol to the Convention. Under the Convention a State becoming a party must accept the regulations dealing with pollution from oil and noxious liquid substances in bulk (which are set out in Annexes I and II respectively), but is given a choice as to whether it accepts the regulations dealing with pollution from harmful substances carried by sea in packaged forms, sewage and garbage (set out in Annexes III-V respectively). The effect of the Protocol is to provide that a State may become a party to the Convention initially by accepting only Annex I (i.e. the oil pollution regulations). Annex II will not become binding until three years after the entry into force of the Protocol or such longer period as may be decided by the parties to the Protocol. This will give States more time to overcome certain technical problems related to Annex II (notably the treatment and disposal of chemical tank washings), which have been partly responsible for the slow rate of ratifications of the Convention. The Protocol also modifies a number of provisions of Annex I, the principal effect of which is to require segregated ballast tanks[3] and/or crude oil washing[4] on both new and existing oil tankers, thus extending the construction standards of the Convention and applying them to existing as well as new tankers. The conditions for the entry into force of the Protocol are the same as for the Convention, and a target date of June 1981 has been set for the entry into force of the Convention as modified by the Protocol.

2.2 Conventions Concerned with Liability for Pollution

2.2.1 The 1969 Oil Pollution Liability Convention (CLC) and the 1971 Fund Convention

The main aim of the CLC is to make it easier for a person who has suffered damage as a result of oil pollution to recover compensation from the shipowner whose vessel was responsible. It achieves this aim by simplifying certain aspects of legal procedure, notably by providing that the shipowner, subject to a number of exceptions, is strictly liable for any damage caused by oil pollution (thus obviating the need for the plaintiff to prove fault or negligence on the part of the shipowner—which in practice would often be difficult): the shipowner's liability, however, is limited unless he has been guilty of fault. The purpose of the Fund Convention is to establish an International Oil Pollution Compensation Fund (financed by contributions from oil companies) which is to provide the victim of oil pollution damage with additional compensation where the damage he has suffered exceeds the shipowner's limits of liability or where the shipowner is not liable at all by virtue of one of the exceptions referred to in the CLC (war, failure of a government to provide lights, etc).

These two Conventions were a direct outcome of the *Torrey Canyon* disaster. They were both drafted by IMCO's Legal Committee and were adopted at conferences, arranged by IMCO, held in Brussels in November 1969 and December 1971. IMCO also acts as depositary for both Conventions. IMCO is further given the task of convening conferences for the purpose of revising or amending the Conventions, and IMCO did in fact convene such a conference in November 1976. This conference amended the two Conventions by expressing the limits of liability in terms of Special Drawing Rights and not in terms of Poincaré francs as hitherto. Under the 1971 Convention IMCO also has the task of convening the first session of the Fund's Assembly (which took place in November 1978) and of being notified by each State party of its companies which are required to contribute to the Fund.

The CLC came into force in June 1975 and the Fund Convention in October 1978. During the considerable period of time which elapsed between the adoption of these Conventions and their entry into force, victims of oil pollution were able to claim compensation under two oil-industry sponsored schemes—known as TOVALOP and CRISTAL, and described in Chapter 4—which roughly parallel the CLC and Fund Convention. The CLC has been ratified by 36 States. Although this may not seem a very high figure, it does in fact represent nearly 85 per cent of the total world tanker tonnage (and it is ratification by tanker-owning States which largely determines the effectiveness of the Convention). The Fund Convention has been ratified by 15 States. Here, however, it is ratification by oil-importing (as opposed to tanker-owning) States which is important, because it is oil importers which finance the Fund. The present 15 States parties are collectively responsible for nearly 50 per cent of total world oil imports.

2.2.2. Liability for Other Forms of Pollution

IMCO's Legal Committee has spent several years discussing whether the CLC should be extended to cover liability for marine pollution from ships by substances other than oil. As yet its discussions are not very far advanced, and there is some division of opinion within the Committee as to the need for such an extension of the 1969 Convention. If the Committee does decide to proceed with drawing up a new Convention, it will no doubt be several years before any such Convention is ready for adoption.

2.3 Conventions Concerned with Coastal State Intervention

2.3.1 The 1969 Intervention Convention

The object of the International Convention relating to Intervention on the High Seas in Cases of Oil Pollution Casualties is to determine the circumstances in which, and the measures, a coastal State can take where its coasts are threatened with oil pollution from a wrecked or stranded ship. Like the CLC, this Convention was a direct outcome of the *Torrey Canyon* disaster, and it, too, was drafted by IMCO's Legal Committee: it was adopted at the same IMCO-sponsored conference as the CLC, and

IMCO acts as depositary for this Convention too. The Convention also confers a number of other tasks upon IMCO. First, the Organisation is to maintain a list of experts whom any contracting State can consult before taking intervention measures under the Convention. Second, when a State has taken such measures, it must notify IMCO. Third, if any dispute between parties to the Convention is submitted to the conciliation or arbitration procedure set out in the Annex to the Convention and the parties cannot agree on the presiding member of the Conciliation Commission or the Chairman of the Arbitration Tribunal, such person shall be appointed by the IMCO Secretary-General. Lastly, IMCO may convene conferences for the purpose of revising or amending the Convention: to date no such conference has been held.

The Convention came into force in May 1975 and currently has 32 contracting parties, including all the leading tanker-owning States with the exception of Greece, Singapore and Italy. On the other hand, ratification by coastal States, particularly by developing States, is rather low: this may be because they think that the Convention does not give them sufficient powers.

2.3.2 The 1973 Intervention Protocol

At the same IMCO Conference at which MARPOL was drawn up, a protocol to the Intervention Convention was also adopted — the Protocol relating to Intervention on the High Seas in Cases of Marine Pollution by Substances Other than Oil. The Protocol, when it enters into force (and so far it has only received 3 of the 15 ratifications necessary), will give States parties the same rights to take action against pollution from substances other than oil as the 1969 Convention gives them in cases of oil pollution. The main function of IMCO under the Protocol (apart from acting as depositary and convening review conferences) is to draw up a list of 'substances other than oil': the MEPC has in fact drawn up such a list.

2.4 Appraisal of the Above Conventions

By comparing the list of Conventions discussed in Chapter 1 with the Conventions outlined above, it can be seen that IMCO has been involved in virtually all the general multilateral Conventions concerned with marine pollution from ships. In the case of the regional and bilateral agreements that exist alongside the general Conventions, IMCO has not been involved until recently. Under the Mediterranean Pollution Convention of 1976, a Regional Oil Combating Centre has been established in Malta. The main responsibility for the establishment and operation of

the Centre has been entrusted to IMCO. The Centre's principal functions
are to facilitate co-operation among the coastal States of the region to
combat massive spillages of oil and to assist the Mediterranean States,
where necessary, in developing their own national capabilities to combat
oil pollution. IMCO is also assisting in the setting up of a similar centre
in the Persian Gulf, further to the 1978 Kuwait Regional Convention for
Co-operation on the Protection of the Marine Environment from
Pollution.

The value of the IMCO Conventions surveyed above has been recognised
by the Third United Nations Conference on the Law of the Sea. The
informal Composite Negotiating Text (ICNT) produced by the Conference
provides that a coastal State may not normally prescribe pollution regulations for shipping in its territorial sea or exclusive economic zone unless
such regulations are in conformity with 'generally accepted international
rules and standards'. The ICNT also provides that States 'shall establish
laws and regulations' for preventing pollution from their vessels; and
'such laws and regulations shall at least have the same effect as that of
generally accepted international rules and standards.' Although the ICNT
nowhere defines such 'generally accepted international rules and standards',
they are generally taken as referring to the rules and standards contained
in the IMCO Conventions referred to above. Thus, if these provisions of
the ICNT are retained in any eventual convention produced by the Conference and such a convention were to enter into force and be widely
ratified, the rules and standards of the IMCO Conventions would be applied
more widely than at present, because they would also be applicable, to the
extent outlined above, to non-parties to the Conventions, and not just to
the parties as at present.

Even though the UN Conference has recognised the value of the IMCO
Conventions, their usefulness and the role of IMCO must not be exaggerated.
The Conventions listed under headings 2 and 3 above do not do anything to
prevent or control marine pollution from ships: they are limited merely to
providing compensation when pollution has occurred or mitigating its
effects in cases of shipping casualties. It may be, of course, that the fact
that it is now easier for a shipowner to be sued for oil pollution damage
may act as some kind of deterrent to his permitting his ship deliberately
to pollute the sea: so far, however, there is little evidence of this. While
the Conventions reviewed under heading 1 are directly concerned with
preventing or controlling pollution by setting construction and discharge
standards and by prohibiting certain kinds of discharge altogether, the
role of IMCO is a strictly limited one. Although IMCO sets pollution
standards through its Conventions, its role ends there. It has no say in

how its Conventions are implemented or enforced: this is a matter entirely in the hands of the States which are parties to the various Conventions.

More far-reaching criticisms of IMCO marine pollution Conventions and the role of IMCO have been made, however, and these must now be examined. First, it is said that IMCO Conventions are slow to enter into force. There is no denying the force of this criticism: of the Conventions discussed above which have come into force, the period of time between the adoption and the entry into force of each Convention was on average five years—during which time of course pollution of the sea continued. Moreover, the most important Convention of all—MARPOL—is still a long way from coming intó force. Unfortunately in international law, delay in the entry into force of Conventions is not confined to IMCO Conventions or even marine pollution Conventions: it is a general feature of the international legal system. There is little that IMCO itself can do to remedy this situation. Entry into force of Conventions requires that the stipulated number of ratifications be attained, and ratification is a matter completely in the hands of States. Whether a State ratifies a particular Convention, and, if it does, the speed at which it does so, are matters completely within its discretion. Its actions—or inactions—will be guided not only by its view of the Convention in question and the political pressures which may be brought to bear by particular national interest groups, such as shipowners, who are directly affected by the Convention, but also by such practical considerations as the need to find time in the national legislature for approval of ratification (if such approval is required) or for the passage of any implementing legislation that may be necessary. Additional factors guiding a State's actions will be the degree of technical development required and the economic implications of becoming a party to any particular Convention: where both such factors are far-reaching, as in the case of MARPOL, it is inevitable that ratification will be slow. In this situation there is little that IMCO can do except encourage its member States which have not done so to ratify the various Conventions as speedily as possible, and in fact at its recent sessions the IMCO Assembly has adopted resolutions urging such ratifications. This process was taken a step further by the Council in 1978, when it appointed three consultants to visit member States and explain the outcome of the 1978 Conference on Tanker Safety and Pollution Prevention: in addition, it also appointed two permanent senior officials to promote the implementation of IMCO Conventions generally. As far as Amendments to Conventions are concerned, the adoption of the tacit amendment procedure (which was described above) should mean

that Amendments to the technical parts of Conventions will in future
enter into force more rapidly than has been the case in the past.

A second criticism is that IMCO Conventions, even where they have
come into force, have not attracted many ratifications. At first sight there
appears to be some force in this criticism, since in relation to the five
Conventions discussed earlier which have come into force, the number
of ratifying States ranges from 15 to 59: this contrasts with the fact that
the total number of States in the world is now in excess of 150. However,
as has already been hinted at earlier, mere numbers are somewhat misleading. The important question is not so much how many States have
ratified, as which States. There is little point, for example, in a Convention designed to prevent oil pollution from oil tankers being ratified
by a large number of non-tanker-owning States but by few or no tanker-
owning States.[5] From this point of view, the IMCO Conventions have
been more successful than the number of their ratifications might at first
suggest. As has already been pointed out, the 1954 and the two 1969
Conventions have been ratified by States which collectively own 80 per
cent or more of the world's oil tankers, at which these Conventions are
principally aimed. A further point one should not overlook is that sometimes when a State feels it is not able to ratify a particular Convention,
it does nevertheless apply some or all of the Convention's provisions to
its shipping. As suggested above, this phenomenon may become much
more common should the present UN Conference on the Law of the
Sea's negotiating text become a widely ratified Convention.

Third, a criticism sometimes made by environmentalists is that the
IMCO Conventions dealing with deliberate pollution (those dealt with
under heading 2.1 above) are too industry-oriented in the discharge and
construction standards they set. There is not space in this chapter to
consider this criticism in relation to specific standards laid down by the
various Conventions, but one or two general points can be made. There
is little point in a Convention laying down standards for the implementation of which the necessary technology does not exist or is regarded as
unduly costly, for such a Convention will simply not be ratified and
brought into force. It is the fact that some of the standards laid down in
MARPOL are at the limits of current technical knowledge that helps to
explain why that Convention has so far attracted so few ratifications.
What any Convention has to do is to try and strike a balance between
aiming at the best possible protection of the environment and the setting
of discharge and/or construction standards which it is both technically
possible and not financially crippling for the shipping industry to comply
with. If MARPOL, as modified by its 1978 Protocol, does come into

force in the fairly near future, there will be some cause for saying that IMCO has struck about the right balance here.

Lastly, it is often said that the IMCO Conventions look fine on paper, but in practice they are not properly implemented and enforced. There is probably a good deal of truth in this. Here again, IMCO itself is relatively impotent. It is up to the contracting States to implement and enforce the Conventions. As far as enforcement is concerned, under the IMCO Conventions (subject to some minor exceptions) this rests solely with the flag State.[6] There is a good deal of evidence that not all flag States are able or willing to enforce the Conventions as effectively as they are in theory required to do. Had the procedures under the 1954 Convention for reporting to IMCO on implementation and enforcement of that Convention been more widely observed, there seems little doubt that IMCO could have brought some pressure to bear on recalcitrant States. One can only hope that the equivalent procedures in MARPOL will be better observed if and when that Convention enters into force. In this connection it is interesting to note that one of the items that the IMCO Assembly in December 1977 placed on the long-term work programme of the Marine Environment Protection Committee was the development of suitable procedures for the enforcement of Conventions relating to marine pollution, including the question of penalties. The change of emphasis in marine pollution control from discharge to construction standards will also help to make enforcement more effective: it is much easier to see whether a ship is constructed in accordance with the relevant standards than it is to determine whether a particular discharge exceeds the permitted level.

3 CONVENTIONS INDIRECTLY HELPING TO PREVENT MARINE POLLUTION

In the previous section we discussed the various IMCO Conventions which are *directly* concerned with marine pollution. In this section we must consider a number of Conventions drafted and adopted by IMCO which, by promoting safety at sea, help to reduce marine pollution occurring as a result of shipping casualties and accidents, such as hull failure, collisions and strandings. It has been calculated that such accidents account for about 12 per cent of all oil pollution of the sea originating from ships.

There are six Conventions to be considered, many of them of considerable complexity. For reasons of space, however, their provisions can only be very briefly summarised here.

3.1 International Conventions for the Safety of Life at Sea 1960 and 1974 (SOLAS Conventions)

There has been a succession of SOLAS Conventions, the first of which was prompted by the sinking of the *Titanic*. The current Convention is that of 1960, which is due to be replaced by that of 1974 when the latter has received sufficient ratifications for its entry into force. The SOLAS Conventions are of considerable bulk and complexity, each running to nearly 250 pages of text and containing over 200 regulations. The chief regulations of interest, as far as the prevention of accidents is concerned, are those dealing with the construction of ships, the mandatory carriage of modern electronic navigational equipment and other aspects of the safety of navigation, and the carriage of dangerous goods. SOLAS 1960 has received 99 ratifications, representing over 95 per cent of the total world merchant shipping fleet, while SOLAS 1974 is expected to meet the conditions necessary for its entry into force in the fairly near future.

In 1978, at the same IMCO Conference on Tanker Safety and Pollution Prevention at which the Protocol to MARPOL was adopted, a Protocol to SOLAS 1974 was also adopted. This Protocol will make the use of inert gas systems,[7] additional radar and emergency steering gear mandatory on all ships above a certain size, and will improve procedures for the inspection and certification of ships.

3.2 Convention on the International Regulations for Preventing Collisions at Sea 1972

Attached to this Convention is a set of regulations prescribing conduct which, when observed, should prevent collisions between ships. Of particular interest is Rule 10, which makes the observance of traffic separation schemes[8] mandatory. The SOLAS Conventions provide that IMCO is the only international body competent to prescribe such schemes. IMCO has in fact been prescribing such schemes for over a decade now, although before the 1972 Convention entered into force (in mid-1977) the observance of such schemes was entirely voluntary. Currently there are over 70 IMCO prescribed schemes in existence. The 1972 Convention has been ratified by 55 States, representing about 85 per cent of the total world merchant shipping fleet.

3.3 International Convention on Load Lines 1966

Overloading is often the cause of casualties to ships, particularly cargo ships. The 1966 Load Lines Convention prescribes the minimum freeboard (or maximum draught) to which the ship is permitted to be loaded. The

Convention has been ratified by 91 states, representing over 95 per cent of the total world merchant shipping fleet.

3.4 Convention on the International Maritime Satellite Organisation 1976

The Convention provides for a world-wide maritime satellite system (replacing the present two US Marisat satellites and the European Marots satellite) which, when it begins operations in the early 1980s, should lead to a significant increase in the speed, reliability and quality of maritime communications. This in turn will make for increased efficiency of navigation and safety at sea.

3.5 International Convention on Standards of Training, Watchkeeping and Certification of Seafarers 1978

Studies have shown that inadequately trained seamen, who are found particularly on flags of convenience, are a major factor in the causes of shipping accidents. The Convention establishes mandatory minimum requirements for the training, certification and watchkeeping of seamen. Thus when the Convention comes into force, it should help to raise standards of seamanship, thereby leading to a general increase in the safety of navigation.

3.6 Convention Concerning Minimum Standards in Merchant Ships 1976

This Convention was in fact adopted under the auspices of the International Labour Organisation, and not IMCO. Nevertheless IMCO played an important part in its drafting. The Convention attempts to deal with the problem of substandard ships, which are most commonly associated with flags of convenience. It provides that States parties must apply to their ships and enforce certain minimum safety standards, social security measures and shipboard conditions of employment. They must also ensure that seafarers on their ships are properly qualified and they must hold official inquiries into serious marine casualties involving their ships. The Convention also confers a limited form of jurisdiction on port States (going beyond the powers of control such States have under the SOLAS and Load Line Conventions). It provides that where a State finds that a ship in one of its ports does not conform to the standards of the Convention, it may make a report to the flag State and the ILO, and may also 'take measures necessary to rectify any conditions on board which are clearly hazardous to safety or health', provided that such measures do not involve unreasonable detention or delay. The ILO Convention ties in with the IMCO Conventions by providing that States may only become parties to it if they have ratified the IMCO Conventions referred to in 1-3 above.

3.7 Appraisal of the Above Conventions

SOLAS 1960, the Load Lines Convention and the 1972 Collisions Regulations Convention have made a significant contribution to the promotion of safety at sea and thus to the avoidance of accidental pollution from shipping. An even greater contribution will be made by the other Conventions, particularly SOLAS 1974 and its 1978 Protocol, if and when they enter into force and are widely ratified.

4 OTHER MEASURES TO COMBAT MARINE POLLUTION

So far we have discussed IMCO's work relating to marine pollution in terms of legally binding conventions, but a good deal of its work in this area takes other forms—notably recommended codes of practice, technical assistance, research and education. Each of these will now be examined briefly.

4.1 Recommended Codes of Practice, Guidelines etc.

Coming under this heading are such things as the International Maritime Dangerous Goods Code, Codes for the Construction and Equipment of Ships carrying Dangerous Chemicals in Bulk (1971) and of Ships carrying Liquified Gases in Bulk (1975), a Recommendation on International Effluent Standards and Guidelines for Performance Tests for Sewage Treatment Plants, a Recommendation on International Performance and Test Specifications for Oily Water Separating Equipment and Oil Content Meters (1977) and Guidelines on the Provision of Adequate Reception Facilities in Ports. These recommended codes of practice are generally adopted in the form of resolutions of the IMCO Assembly and therefore are not as such legally binding. Nevertheless they are quite widely observed and some States have even incorporated these codes into their national legislation: for example, the Dangerous Goods Code has been incorporated into the national legislation of 20 States and partially incorporated in a further 7 States.

4.2 Technical Assistance

Since 1965 IMCO has operated what is now a rapidly growing programme to provide technical assistance in shipping matters to developing countries. This assistance chiefly takes the form of providing experts to help with particular projects (which have included projects relating to marine pollution). The programme is funded, not by IMCO itself, but by the UN Development Programme, bilateral aid agencies and the recipient countries themselves. The programme was considerably strengthened in 1978 when, as a result of the Conference on Tanker Safety and Pollution Prevention,

IMCO established a Marine Safety Corps. The Corps is designed to provide a nucleus of experts who will assist governments (particularly those of developing countries) in formulating regulations and carrying out surveys and inspections of ships in accordance with international conventions relating to safety and pollution prevention.

4.3 Research

IMCO has neither the personnel nor the facilities for carrying out any large-scale research into the scientific, technical, economic or legal aspects of marine pollution from ships. IMCO does, however, take part in the work of, and provide the administrative centre for, the Group of Experts on the Scientific Aspects of Marine Pollution, a body which advises those UN specialised agencies whose work involves questions of marine pollution. IMCO has also on occasions commissioned research from outside bodies: for example, in preparation for the conference which adopted MARPOL in 1973, IMCO commissioned nine background studies from member governments on the economic aspects of the proposed Convention. Some of these studies, however, have been criticised as showing some bias towards ship-owning interests.

4.4 Educational Activities

IMCO engages in a number of activities aimed at educating the national authorities of member States in questions of marine pollution. Thus, it held a symposium on various aspects of MARPOL in Mexico in 1976; it is currently organising a series of seminars on the results of the 1978 Tanker Safety and Pollution Prevention Conference in London, Asia, Africa and Latin America; courses on the implementation of marine pollution Conventions were held in Sweden in 1977 and 1978; and a workshop on marine pollution was held jointly with the UN Environment Programme in West Africa in 1977. Also as part of its educational activities, IMCO has published a *Manual on Oil Pollution* which deals with methods of preventing oil pollution from ships, contingency planning, salvage of oil from stricken vessels, practical information on means of dealing with oil spillages and legal aspects.

5 IMCO'S FUTURE WORK ON MARINE POLLUTION

In December 1977 the IMCO Assembly adopted the Organisation's Long Term Work Programme up to 1984. In the field of marine pollution, the principal objectives of the Programme are solution of the technical problems involved in implementing MARPOL, the development of suitable procedures for enforcing marine pollution conventions and the

promotion of technical co-operation relating to marine pollution (including regional arrangements for combating pollution). In addition, a number
of specific tasks for the MEPC were spelt out. These include promotion
of reception facilities for residues and oil discharge monitoring systems,
identification of the source of discharged oil, production of a manual on
intervention under the 1969 Convention and its 1973 Protocol,
procedures and arrangements for the discharge of noxious liquid
substances and the prevention of pollution by such substances carried in
bulk, and the inclusion of pollutants in the International Maritime Dangerous Goods Code.

Following the *Amoco Cadiz* disaster[9] in March 1978, the IMCO Council
requested the Legal Committee to work on a number of points not
covered by the above Work Programme. The Committee was asked to consider the existing compensation regime, to study ways of improving the
existing international legislation relating to the salvage and assistance of
ships involved in maritime incidents, and to consider the legal aspects of
plans to make it mandatory for ships which are involved in difficulties
likely to result in pollution to report to nearby coastal States. At the same
time the Council established a special *ad hoc* working group to consider
the relationships between the masters of ships, owners and flag States,
together with the significance of having courts of inquiry in all shipping
countries. The Maritime Safety Committee has also taken action in
response to the *Amoco Cadiz* disaster. It has modified the traffic separation scheme off the Brittany coast where the *Amoco Cadiz* ran aground:
in future laden oil tankers and other ships carrying hazardous cargoes will
have to keep much further from the French coast. The Committee has
also agreed to study ships' steering systems (including duplicated steering
gear and rudders), ways of improving the reliability of ships' components
which are vital for manoeuvring the ship in cases of emergency, and
safety features associated with flags of convenience.

6 CONCLUSIONS

From its modest beginnings, IMCO has today reached a position where
its expertise in technical matters relating to shipping is widely and
deservedly recognised. It has managed to achieve this with relatively
limited manpower and financial resources, and without spending a lot of
its time and energy on emotive but irrelevant political issues such as the
Middle East and Southern Africa, in the way that has so frustrated the
work of many other UN specialist bodies. As an organisation, IMCO has
deliberately kept a low profile and concentrated almost exclusively on
technical questions. As a result it has built up an impressive body of

conventional law on marine pollution. These Conventions, however, are not without their limitations, but the limitations from which they do suffer—slowness at coming into force and inadequate enforcement—are more the result of weaknesses in the international legal system itself than due to shortcomings in IMCO. It must also be remembered that IMCO deals with only one source of marine pollution—pollution from ships. This is not the most important source of marine pollution—pollution from land-based sources is much more significant—but it is one of the least difficult to control effectively.

BIBLIOGRAPHICAL NOTE

Most of the Conventions referred to in this chapter are discussed in Chapter 1. References to their sources will be found there and in the List of International Instruments. Many of these Conventions are also conveniently reproduced in R. Churchill *et al.* (eds.), *New Directions in the Law of the Sea*, Vols II, IV & VI. IMCO has given an account of its work in *The Activities of IMCO in relation to Shipping and Related Maritime Matters* (1974); this may be supplemented by *IMCO News*, published quarterly. Useful articles on IMCO's work in relation to marine pollution include T. Busha, 'IMCO Conventions on Pollution of Coastal Waters and Harbours' and P.W. Birnie, 'International Standards: Their Present Shortcomings and Possible Development' in J. Nowak (ed.), *Environmental Law, International and Comparative Aspects* (1976), and L. Juda, 'IMCO and the Regulation of Ocean Pollution from Ships', (1977) 26 *I.C.L.Q.* 558.

NOTES

1. In March 1967, the *Torrey Canyon*, a Liberian-registered tanker ran aground on the Seven Stones reef off Land's End, spilling over 100,000 tons of oil into the sea, which caused widespread damage to the British and French coasts and killed a large number of fish, birds and other forms of marine life. At the time it was the most spectacular case of oil pollution resulting from a shipping accident.
2. 'Flags of convenience' refer to the practice whereby, for tax and other financial reasons, a ship is registered in a State with which neither the ship nor its owner has any real connection. As a rule, such flags of convenience States have little real control over or contact with their vessels. The major flags of convenience States are Liberia and Panama, which account for 20 and 5 per cent respectively of the total world merchant shipping fleet.
3. I.e. the tanks used for ballast water are kept separate from the tanks used for carrying the oil cargo. Traditionally, oil tankers have filled their empty oil cargo tanks with sea-water to act as ballast on their return voyages after having unloaded their cargoes. It is when this ballast water is emptied before the tanker loads up with oil cargo again that pollution has often occurred in the past.
4. This is a method of cleaning cargo oil tanks by pumping crude oil from the

cargo into the cargo tanks to be cleaned and spraying the surfaces of the tank *via* fixed tank washing machines. The solvent action of the crude oil has the effect of cleaning the oil from the surface of the tanks. The oil washed from the tanks is pumped out of the vessel with the remainder of the cargo.

5. This point now seems to be receiving general recognition by IMCO. For example, MARPOL with its 1978 Protocol requires a relatively small number of ratifications for its entry into force (15), but the combined merchant fleets of these States must constitute not less than 50 per cent of the gross tonnage of the world's merchant shipping.

6. It has been widely suggested that enforcement of IMCO Conventions would be more effective if coastal States and port States were given some competence in this area. While this view has the support of many IMCO member States, it has been felt by IMCO that the question is one for resolution by the UN Conference on the Law of the Sea rather than by IMCO. The current Conference negotiating text does in fact give jurisdiction to both coastal and port States to enforce pollution conventions, subject to certain conditions and limitations. For further discussion of this question, see Chapter 3 of this book.

7. Such systems are designed to reduce the risk of explosions in empty oil tanks by pumping into the tanks gases so deficient in oxygen that the atmosphere within a tank may be rendered inert, i.e. incapable of propagating flame.

8. A traffic separation scheme separates shipping in congested areas into one-way-only lanes.

9. The *Amoco Cadiz*, a Liberian oil tanker of 220,000 tons, drifted on to the Brittany coast after a failure in its steering gear when the ship was some 8-10 miles off-shore. As a result of its stranding, most of the ship's cargo of 220,000 tons of crude oil was washed up on to the Brittany beaches, causing enormous damage to the tourist industry and local fishermen. It is the worst pollution disaster in history.

3 ENFORCEMENT OF THE INTERNATIONAL LAWS FOR PREVENTION OF OIL POLLUTION FROM VESSELS

Patricia Birnie

1 INTRODUCTION

Pouring oil on to water used to be an acceptable method of preventing trouble: nowadays it is a source of international friction. In United Kingdom waters alone there were, in 1977, 658 oil spills, only 308 of which were traced to the vessel responsible.[1] Few of these could be prosecuted in Britain because either the offending discharge did not occur within UK jurisdiction or insufficient evidence was obtainable while the vessel was in a UK port.

Oil spills occur in spite of the proliferation of Conventions to prevent them, partly because not all the Conventions are in force and because of weaknesses in those that are, but mainly because ratifying States do not always strictly enforce them on their vessels. This laxity is facilitated by the prevailing legal doctrine of freedom of navigation on the high seas coupled with a right of innocent passage for foreign vessels through territorial waters. Restrictions on vessels delaying passage impose extra costs on shipping companies and these economic implications have led States with predominantly shipping interests to resist enforcement by States whose ports or coastal waters vessels enter. The growth of the oil industry, the concomitant development of larger tankers carrying progressively more hazardous cargoes with increasing frequency on the favoured near-coastal routes and through narrow straits has intensified demands for revision of the customary doctrines, yet all States, as consumers of sea-borne goods, are interested in their economical and speedy transport.

As Conventions are not widely enforced if not widely accepted, political compromises have to be made between the environmental interests of coastal States and the commercial interests of shipping States, interests which, as in the UK, may coincide in the same State. Shipping is an international business; enforcement cannot become a patchwork of coastal State regulations. Concessions to coastal State jurisdiction come gradually, by negotiation in international *fora*, subject to qualifications ensuring an equitable balance of interests.

1.1 Freedom of the Seas: the Classical Doctrine

Until recently the enforcement of law on vessels at sea was based on two
simple assumptions: that States were entitled to extend their territorial
sovereignty only over the narrow belt of adjacent waters (the territorial
sea) necessary to protect their coasts from military attack rather than
danger to their marine environment, and that foreign vessels could
exercise a right of 'innocent passage' therein, non-innocence being determined by degree of threat to the coastal State's security, defined in
terms excluding environmental harm. The seas beyond a 3-12 mile
belt were regarded as high seas on which vessels were, if registered in and
flying the flag of a State, free to voyage without interference.
Applicable international law was enforceable only by the flag State,
generally on the vessel's return to it,[2] but there was until after World
War II little law relating to preventing oil pollution. The first treaty on
Prevention of Pollution of the Sea by Oil was negotiated (by diplomatic
conference) only in 1954.[3]

1.2 IMCO

The United Nations, after its establishment in 1945, promoted the
institution in 1958 of IMCO, the work of which has been described in
the preceding chapter. The enforcement provisions of IMCO Conventions
and recommendations have gradually augmented the classical doctrine,
but it must be stressed that IMCO is not a supranational body and has
never been given enforcement powers; some Conventions do allow for
reporting of offenders to IMCO, but it cannot take punitive action against
them. It provides a *forum* for balancing the interests of flag and coastal
States, and is urgently considering revision of coastal State powers following the *Amoco Cadiz* disaster;[4] but, as voting in IMCO is weighted
according to gross registered tonnage (GRT), and as moreover some
Conventions do not enter into force until ratified by a specified number
of States representing a fixed amount of GRT,[5] its Conventions
primarily protect shipping interests. New enforcement provisions do
not bind even ratifying States until a Convention enters into force
(which takes 5 years on average); IMCO's new procedures for tacit amendment and non-binding Target Dates for entry into force could speed this
process up, but are limited in scope and to recent Conventions.[6]

1.3 The Codification of the Freedom of the Seas: UNCLOS I, 1958

1.3.1 The Territorial Sea Convention

In the year of IMCO's establishment the UN also convened its First

Conference on the Law of the Sea (UNCLOS I) which, *inter alia*, adopted a Convention on the Territorial Sea[7] which set no outer limit therefor, but in Articles 14-17 confirmed that the coastal State's sovereignty over it was subject to rights of innocent passage as long as there was no threat to 'the peace, good order and security of the coastal State'. Maritime States resisted any interpretation of this which would allow coastal States to impede the passage of vessels allegedly threatening their marine environment but, even before UNCLOS I and the 1954 Convention, three South American States had, in the 1952 Declaration of Santiago,[8] proclaimed their jurisdiction over a 200-mile 'maritime zone'. Originally directed at resource control, the concept was in time adopted and adapted by other States to include jurisdiction for the purpose of preserving the marine environment, especially from vessel-source pollution. Canada is, however, the only State to date[9] to have unilaterally promulgated, in its 1970 Arctic Waters Pollution Prevention Act,[10] a zone of 100 miles exclusively for pollution prevention, laying down construction and passage requirements. Canada claimed that the ecological fragility of Arctic waters rendered them a 'special area' justifying mandatory coastal State control, but many States protested its illegality. Until UNCLOS III produces a treaty or generates new customs approving such zones, their legal status remains doubtful.

The Territorial Sea Convention limits the coastal State to taking 'the necessary steps' to prevent non-innocent passage therein, though requiring foreign ships to comply with its laws (if they conform to the Convention and to international law) regulating, *inter alia*, transport and navigation. The coastal State's criminal jurisdiction can only be exercised if the crime on board affects the coastal State or disturbs the good order of the territorial sea. No mention is made of pollution in this context. These doctrines clearly require reconsideration in the light of modern ships and the incidence of massive pollution. The Territorial Sea Convention has been ratified by less than a third of the States attending UNCLOS III, many of which also deny that its doctrines are part of customary international law. UNCLOS III, as we shall see, is therefore attempting to develop carefully defined concepts of Exclusive Economic Zones (EEZ), special areas and acts of non-innocent passage which will strengthen but limit the powers of coastal States.

1.3.2 The High Seas Convention: Substandard Ships and Flags of Convenience

The High Seas Convention[11] codified freedom of navigation thereon; ships remain subject only to the jurisdiction of their State of registration, whose flag they fly. The flag State must have a 'genuine link' with the vessel but no criteria are laid down; the link is often tenuous. The *Torrey Canyon*, for example, flew a Liberian flag, was built in Japan, registered by a company in the Bahamas, had an Italian master and crew and was on single charter to British Petroleum from the Gulf to Milford Haven.

Some registering States have more stringent requirements for crew and equipment standards,[12] resulting in higher costs; a United States flag costs 30 per cent more than others.[13] Owners cannot always comply with particular national requirements and it is common practice to register vessels under more convenient regulations – hence the term 'flags of convenience'.[14] It does not follow that a vessel so registered is substandard or subject to poor enforcement, though this may be the case. Substandard ships can be found under any flag. Some flags do have more accidents than others, but more vessels are registered under some flags than others.[15] The average loss rate by flag from 1967-76 showed that Cyprus, Somalia, Panama, Singapore, Greece, Spain, Liberia (in that order) had above-average losses.[16]

Much is being done through IMCO and International Labour Organ-isation (I.L.O.) Conventions to improve flag State registration requirements and enforcement. Flag States retain their primacy in enforcement; other States obtaining evidence of violations can generally only report the offending vessel to its flag State (which may not prosecute or even report on its actions), and usually cannot act until the vessel returns to its jurisdiction. Other means of enforcement are now being pursued in IMCO, I.L.O., UNCLOS III, the EEC and in other regional bodies. Progress is limited by the international nature of shipping; international law requires that vessels entering foreign ports must not be discriminated against, and there is a general presumption that they cannot be denied entry,[17] which also restricts development of regional enforcement schemes.

Article 24 of the High Seas Convention required States to 'draw up regulations to prevent pollution of the seas by the discharge of oil from ships, taking account of existing treaty provisions'. In 1958 there were only two relevant Conventions (SOLAS 1948 and the 1954 Convention), but a large number of measures have since been developed to improve enforcement using a variety of techniques ranging from extensions of port and coastal State jurisdiction to detailed specification of flag State obligations. Both the *lex lata* and the proposals *de lege ferenda* concerning enforcement of pollution prevention Conventions have grown recently, and it is important that not only Conventions directly relevant should be enforced, but that a comprehensive regime regulating all aspects of tanker safety should be strictly applied; enforcement of safety and training

regulations concentrates the minds of seafarers, whose human errors
are so frequently the cause of accidents (*vide* the strandings of the *Torrey
Canyon* and *Amoco Cadiz*),[18] on observation of law. The General
Manager of Milford Haven has stressed that the low incidence of oil pollution in that port accrues from its reputation for tight enforcement.[19]

2 LIMITATION OF VESSELS' FREEDOM TO POLLUTE THE SEAS

2.1 Enforcement of IMCO Conventions

Enforcement of the growing number of IMCO Conventions is primarily
left to the flag State but slowly, by a series of carefully negotiated severely
limited measures, Conventions are introducing other means of enforcement,
generally by the port State but, in some cases, by the coastal State.

2.1.1 Conventions Preventing Discharges

*The International Convention for the Prevention of Pollution of the Sea
by Oil 1954, as Amended in 1962, 1969, 1971*

Prohibited zones and oil record books. This Convention, which prohibits deliberate discharge of certain specified kinds and concentrations of
oily mixtures in designated 50-mile 'Prohibited Zones' (extended in 1962
to 100 miles) is left to flag States to enforce, with the important exception
that the vessels of the 60 States party must carry oil record books, which
can be inspected in port. This was an important innovation which has
become one of the building blocks of an improved enforcement regime,
even though it is strictly limited to inspection in port; vessels cannot be
inspected at sea for the purpose of verifying whether they have discharged
oil and, even if the port inspection reveals that they have done so, the
violation cannot be prosecuted by the port State, even if the offence
occurred in its waters; nor can the offence be pursued by any other State
in whose waters it occurred. The offending vessel can only be reported to
its flag State or to IMCO. The latter can do little more than publicise the
offence and, under this Convention, there are no international means of
finding out whether the flag State has punished the offender. Diplomatic
channels are seldom able to obtain such information. The 1969 and
1971 Amendments to the 1954 Convention did nothing to improve enforcement, the former being limited to enabling the use of the Load-on-
Top system for cleaning ballast tanks and the latter relating to
protective tank arrangements and limitation of tank size.

It is violation of this Convention which causes the constant fouling
of beaches, nets, birds etc., and gives rise to international dissatisfaction

with present enforcement methods. Considerable improvement in
enforcement was, however, introduced in 1973.

*International Convention for the Prevention of Pollution from Ships
(MARPOL) 1973.*[20] MARPOL is not in force and is not likely to be so
until the 1980s because its technical requirements present difficulties
for many States. It aims to prevent all forms of pollution from ships,
not just oil pollution, by regulating all the technical aspects of disposal
of all kinds of pollutants listed in 5 Annexes to the Convention. States
will become strictly liable for discharges in violation of the Convention,
without proof of negligence or intent. Originally, a ratifying State had to
accept Annexes I and II in order to ratify the Convention, but the
technical problems posed by Annex II concerning 'noxious liquid
substances carried in bulk' were such that many potential ratifiers, such
as the UK, were deterred from accepting the Convention, because they
did not have the technical means for compliance. In 1978 acceptance
of Annex II was deleted from the ratification requirement,[21] thus
enabling more States to ratify.[22] The Convention also has two
Protocols, relating to reporting incidents involving harmful substances
and providing for arbitration respectively, and its technical requirements are subject to the new IMCO Tacit Acceptance Procedures for
amendments. When the Convention does enter into force, it will
introduce a number of important procedures which will enable tighten-
ing up of the enforcement regime. Pending its entry into force, some
shipping companies are voluntarily enforcing some measures nationally
(detailed in a 'Pollution Prevention Code'),[23] but it is of course the
provisions for improved enforcement on foreign vessels that are the
most important aspects of this Convention. These cannot be imposed,
however, until the Convention is in force.

Annex I, which is the only compulsory Annex, *inter alia* maintains,
with minor adjustments, the oil discharge criteria of the amended 1954
Convention (which it will replace only as and when States ratify the new
one). It requires use of Load-on-Top, carriage of appropriate equipment
for monitoring discharges, and lays down construction requirements,
including provision of segregated ballast tanks (SBT) for new tankers
over 70,000 dwt. These provisions were amended (even though the Convention is not in force) in 1978 by a Protocol laying down new requirements for surveys, certificates, SBT and Crude Oil Washing (COW) which
had been pressed on IMCO by President Carter following the *Argo
Merchant* stranding off the USA, which led to public demand for
improved measures and enforcement thereof.[24]

Special areas: Annex I also introduced, in a limited form, the concept of 'special areas', that is areas which, because of their special vulnerability to pollution (in this case oil pollution), must be protected from all discharges of oil, with minor exceptions. These areas are much more narrowly defined in MARPOL than the 'special areas' proposed in the Informal Composite Negotiating Text (ICNT) produced by the UNCLOS III, as will be seen; they are limited in MARPOL to a few named areas, viz. the Mediterranean Sea, the Black Sea, the Baltic Sea, the Red Sea and 'Gulfs' Area. The North Sea, virtually a completely prohibited zone for the purpose of discharge under the 1954 Convention (as amended in 1969), is not a special area under MARPOL.

Coastal State jurisdiction: The means of enforcement of the measures under this Convention, although giving the flag State the customary primary role, also give both the coastal and the port State carefully limited powers. The flag State must legislate for the necessary prohibitions and sanctions, but other States detecting violations can inform the Administration of the flag State which, if it is satisfied that sufficient evidence is available to enable proceedings to be brought in respect of the alleged violation, should initiate them as soon as possible, promptly notifying the informing party of the action taken.

The Convention prohibits violations 'within the jurisdiction' of States parties, without defining the extent of this jurisdiction, which it was thought in 1973 would soon be settled by UNCLOS III, which was about to hold its first session. It was therefore necessary to study the proposals in the ICNT to ascertain the present state of negotiations concerning the extent of coastal State jurisdiction over vessel-source pollution both in the 12-mile territorial sea and the 200-mile EEZs proposed therein. The MARPOL provision concerning jurisdiction was inserted at a time when the classical doctrines of high seas freedom and innocent passage were under pressure from the South American declarations of coastal State rights in a 'patrimonial sea' and from Canada's unilateral Arctic waters zone. The majority of States at IMCO were anxious to resist such claims by making more moderate concessions to coastal and port State jurisdiction. Within whatever limits for the territorial sea and EEZ that might be agreed at UNCLOS III, or developed by customary law during or after that conference, MARPOL allows the coastal State within whose jurisdiction an offence has occurred *either* to take proceedings against the offender under its national law *or* to transmit the evidence to the flag State for action as above. The coastal State's precise powers to inspect and arrest vessels for these purposes are, however, still the subject of intense discussion and some disagreement at UNCLOS. Penalties for offences are not fixed numerically in MARPOL, but are required (as in the UNCLOS text) to be 'adequate in severity to discourage violations' and to be 'equally severe irrespective of where the violations occur'.

Port State jurisdiction: It is in its provisions for port State inspection (Articles 5 and 6) that MARPOL breaks the newest ground in enforcement procedures. States into whose port a vessel enters are given some enforcement powers, but these powers are very strictly limited by the terms of the Convention and are less wide in scope than those proposed by UNCLOS III in Article 219 of its ICNT.[25]

Inspection of certificates: Port States are restricted to three activities— verification of the certificates required by the Convention; inspection for illegal discharges; detention of substandard and other ships. Vessels registered in States parties must carry certificates relating to observance of MARPOL's construction and equipment standards. The certificates are issued by the registering State and are then valid in all States parties, the authorised officials of which can inspect them when the vessel is in their port. Inspection for this purpose is generally limited to verification of the certificate. Ships can, however, be inspected to monitor observance of MARPOL's discharge provisions and, since the procedure to be followed is not specified, it could be interpreted widely to obviate the limitations of the certificate inspection. If evidence of violation or the discharge requirements is obtained, it must be forwarded to the flag State which, if it finds the evidence sufficient, must take the proceedings already described. This discharge inspection can also be carried out at the ports and off-shore terminals of States parties on the request of another State party which submits to them 'sufficient evidence' of a violation occurring in 'any place' (presumably within their jurisdiction), but States parties are not obliged to act on such requests and, if they do so, must limit inspection to seeking evidence of the unlawful discharge, reporting the outcome to both the requesting State and the flag State.

Detention of vessels: Ships can only be detained if 'there are clear grounds for believing that the condition of the ship or its equipment does not correspond with the particulars of that certificate', or the ship does not carry a valid certificate. The inspecting party is limited to taking such steps as will ensure that the ship shall not sail until it can proceed to sea without presenting an unreasonable threat of harm to the marine environment. Though the party concerned can permit it to leave the port or off-shore terminal for the purpose of proceeding to the nearest appropriate repair yard available, it is not obliged to do so. There are difficulties in exercising port State jurisdiction because vessels frequently do not spend more than a few hours in port; a proposal to extend this provision by requiring States parties to inform the next port of call of any suspected non-compliance should do much to improve this method of enforcement.

Application to non-parties: Perhaps the most interesting innovation introduced by MARPOL, however, is the requirement that parties *must* apply the *'requirements'* to vessels of States *not* party to it in order 'to ensure that no more favourable treatment is given to such ships' than to those of the parties. At first sight it looks as though this might contravene the customary international law that only the flag State can enforce the international standards on its vessels unless it specifically consents to enforcement by port States, but it should be noted that it is not the provisions of the Convention as a whole that are to be applied but the less formal 'requirements'. This leaves an ambiguity in its application which might be taken advantage of by States wanting to 'ginger up' the standards of States not party to the Convention by conducting informal inspections for such purposes as ascertaining whether the vessel has a certificate etc.; but it seems unlikely that substandard ships of non-parties could be detained or refused port entry unless this can be justified under customary law, i.e. not as provided under this treaty alone. States are beginning to take advantage of this provision as we shall see, even though the Convention is not in force; but they are proceeding cautiously.

2.1.2 Conventions Preventing Collisions and Strandings

Most vessel-source oil pollution of the sea comes from the frequent deliberate discharges which the Conventions referred to above are designed to prevent, and to punish if they do occur. But collisions and strandings of vessels, though less frequent, can, if they do occur and involve supertankers, give rise to massive pollution, as the recent disasters have amply demonstrated. There are a number of Conventions, effective enforcement of which can prevent or reduce such occurrences. These Conventions fall into three categories—those laying down regulations to prevent collisions; those concerning vessel safety; and those eliminating as far as possible the human error which causes many strandings. IMCO has promoted Conventions and elaborated enforcement measures on all these subjects. In addition the I.L.O.[26] has adopted a large number of Conventions specifically relating to the terms, conditions,

training and qualifications of masters and crews on merchant vessels. Few
of these are in force but, if ratified and applied, they would contribute
to the tight regime that would ultimately reduce the present number of
disasters. Meanwhile, some States are beginning to apply the provisions
of some of the Conventions relating to administrative measures, though
the Conventions are not in force, in order to increase the pressure on substandard ships. IMCO Conventions concerning Liability and Compensation
for Oil Pollution Damage and the Establishing of a Fund for the latter
purpose, described in Chapter 1, and now in force,[27] also contribute indirectly to the enforcement regime, as does the knowledge that if a
disaster occurs outside the coastal jurisdiction of States they can, in both
customary law and under the more detailed IMCO Intervention
Convention,[28] take whatever measures are necessary to protect their
coasts, including destroying the stranded vessels.

*Convention on the International Regulations for Preventing Collisions at
Sea, 1966 and 1972.* Both are in force, and their strict enforcement is vital
to prevention of disasters arising from this cause. However, they are
enforceable only by the flag States party to them and by coastal States for
offences occurring in their territorial sea which in some cases, as for example
the UK, is still limited to three miles. Coastal States are, however, now
going to considerable lengths to punish offenders, as the recent prosecution in
a Scottish court of a vessel that had transgressed these regulations in the
English Channel illustrates.[29] Offences in the Channel have reduced dramatic-
ally since the 1972 Convention enabling enforcement of traffic separation
schemes and surveillance improvements.[30]

The International Convention for the Safety of Life at Sea (SOLAS) has
been revised six times; SOLAS 1960 is in force, having been ratified by 98
States, but the most recent SOLAS 1974, which incorporates numerous
Amendments to take account of modern technological advances, is not.[31]
It requires 25 ratifications representing 25 per cent of world tonnage to
enter into force, but now has only 13. Contracting governments of
SOLAS 1974 undertake to give effect to the Convention and apply it on
their flag ships and to deposit with IMCO details of their appropriate laws
and administrative agencies, as well as specimens of the various certificates,
required by the Convention, which verify that their vessels comply with its
requirements. Port States can inspect these certificates, as well as the
MARPOL ones, but only to verify the existence on board of an appropriate
and valid certificate. An inspector can detain a ship until it can proceed
without danger to passengers or crew, but only if there are 'clear grounds
for believing that the condition of the ship or of its equipment does not

correspond substantially with the particulars of that certificate'. Casualties must be investigated by the flag State concerned if it decides that such an inquiry might lead to desirable revisions of the Regulations. Investigating States must report their findings to IMCO, which can issue reports or recommendations based on this information, but cannot reveal either the nationality or the identity of the ship concerned, or impute responsibility for the accident. As certain privileges conceded by the Convention can only be claimed by contracting States, there is some incentive to ratify this Convention, but as yet it is enforceable only by the 13 States that have ratified, and then only on their own flag ships.

The International Load Lines Convention, 1966. [32] This Convention has similar enforcement provisions. It applies, with some exceptions, to flag ships of contracting governments which undertake to effect it, inspecting the validity of the required certificates when the vessel enters port and checking that the vessel is loaded in accordance with its terms.

Convention on Standards of Training, Certification and Watchkeeping for Seafarers, 1978. Conventions relevant to crew training and standards have mainly been developed by the I.L.O. but, because of the greater acceptability to shipowners and governments of the kind of forum provided by IMCO which is more protective of their interests, in 1978 this important Convention was concluded there. It has not yet achieved the necessary acceptance by 23 States owning 50 per cent of world gross tonnage for entry into force but, when it has, it will be applied by contracting parties on their flag ships and by inspection in their ports to ensure that seafarers required by the Convention to be certificated are so certificated, though a certificate must be accepted unless 'there are clear grounds for believing that' it has been fraudulently obtained. As there are frequent allegations that many seafarers carry forged certificates, enforcement of this provision, when the Convention enters into force, should result in a major improvement. An inspecting official who finds a violation must inform either the master or the flag State's Consul so that they can take the appropriate action; only if the violation is not remedied and results in a 'danger to persons, property or the environment' can the ship be detained until this situation has been remedied. Undue delay or detention will give rise to an obligation to pay compensation, so the port State is likely to proceed with caution, but it can, significantly, apply this Article 'as may be necessary to ensure that no more favourable treatment is given to ships entitled to fly the flag of a non-party' than is given to the flag ships of parties.

2.2 Enforcement of I.L.O. Conventions

Long before IMCO tackled the basic problem of eliminating or minimis-
ing the human frailties on board ship that lead to accidents, the I.L.O.
had been tackling this problem through its Joint Maritime Commission,
consisting of equal numbers of seafarers and shipowners, which it
established in 1920 and which over the years has adopted a series of 32
Conventions and 25 Recommendations which, taken together, compose
the International Seafarers' Code, setting standards for almost all
aspects of conditions of life and work at sea. The I.L.O. has particularly
concerned itself since 1933 with eliminating substandard ships and use
of flags of convenience. Few of these Conventions actually enter into
force, for they are subject to ratification often on a variety of conditions
(such as that a named number of States, representing a specified amount
of GRT, ratify), as are the IMCO ones, although the I.L.O., a tripartite
organisation on the organs of which employers, workers and governments
are equally represented, has special procedures for keeping its Conventions in the eye of governments. Recent Conventions impinging on
the reduction of human error include the Conventions on Wages, Hours
of Work and Manning (Sea), 1958;[33] Crew Accommodation (1970);[34]
Concerning Continuity of Employment of Seafarers (1976);[35] Minimum
Standards in Merchant Ships (1976).[36] All are primarily enforceable by
the flag State concerned, which must enact the necessary laws and
regulations, but the 1958 Convention on Wages etc. can also be effected
(with some exceptions) by collective agreements, or by combinations of
the two systems, between shipowners and seafarers. The 1976 Convention
on Continuity of Employment is effected by flag State enactments, in so
far as not affected by collective agreements, arbitration awards or other
national practices.

The 1976 Convention on Minimum Standards, which has only three
ratifications and is not yet in force, is the Convention most relevant to
prevention of casualties. It was adopted following an I.L.O. Report on
'Sub-standard Vessels, Particularly Those Registered under Flags of
Convenience'.[37] This Convention applies to all cargo and other commercial ships, and to 'sea-going' passenger ships, and has no conditions
for entry into force. It prescribes various safety standards including crew
competency, hours of work, manning and safety of life on board, which
ratifying States undertake to enact into their national laws, effecting
these standards on their flag ships. But there are interesting additional
provisions: parties guarantee to exercise effective jurisdiction or control
on their own vessels and also undertake even if they do not have
jurisdiction over a vessel, to satisfy themselves that measures for

effective control of conditions on board other ships are agreed between shipowners' and seafarers' organisations in accordance with I.L.O. Conventions; and that there are procedures for investigation of complaints concerning hiring in the territory of an I.L.O. member State of seafarers of its own nationality on ships registered in a foreign country, any complaints being reported both to the flag State concerned and to the I.L.O., thus internationally publicising the deficiencies. A member party must also ensure that seafarers employed on its flag ships are properly qualified or trained for their contracted duties in accordance with the relevant I.L.O. Recommendations, and must conduct inspections etc. to verify that its flag ships comply with applicable I.L.O. Conventions in force and ratified by it. It must also conduct official inquiries into 'serious' marine casualties involving its flag ships, publishing the final report.

Finally, but perhaps in the future most importantly, a limited form of port State jurisdiction is provided — ratifying States can, if a vessel, even of a non-party State, enters their port, report it to its flag State and to the I.L.O., but only if they receive a complaint that the vessel does not conform to the Convention's standards. They can then take the measures necessary to put right any conditions on board 'that are clearly hazardous to safety or health', as long as they do not unreasonably detain or delay the vessel and also notify the nearest official representative of its flag State. The complaint can come from the crew, a professional body, association or trade union or anyone with an interest in the ship's or the crew's safety, or the health of the latter. Moreover, I.L.O. members are required to advise their nationals, if possible, on the problems of signing on ships of flag States which have not ratified the Convention, unless that State applies the conventional standards.

This Convention is clearly innovatory in a number of ways, especially in its endorsement of administrative enforcement action against the substandard ships of States which are not party to the Convention. This provision is a powerful incentive to ratification for States which want to eliminate substandard vessels from their waters since, by use of the mild harassment implicit in these powers, they can cause States registering such ships considerable public and private discomfort.

3 REGIONAL ACTION

3.1 The Hague Memorandum 1978[38]

As the more recent Conventions — the very ones which improve enforcement by augmenting the primary powers of the flag State by conceding important but nicely qualified powers to the port State — are not in force, following the recent serious casualties, especially the stranding of the *Amoco Cadiz*, which threatened the coasts of both France and the UK, States in Europe began to consider ways in which they could tighten up enforcement of existing international customary and conventional standards on a regional basis, without undermining the continued reasonable freedom of shipping as sanctioned by international law. Eight States bordering the North Sea decided to co-operate in an informal regional arrangement, effected and evidenced not by a treaty or even an inter-governmental agreement, but by a Memorandum of Understanding signed by representatives of relevant concerned Departments in each State party. This so-called Hague Memorandum is not, therefore, a unilateral declaration *de lege ferenda*; it lays down no new standards, but states the intention of the parties stringently and vigorously to exercise the powers which they consider that they already have under certain specified international instruments to the extent that their laws and existing competence permits. They agree to maintain, within their jurisdiction (at present limited to their territorial seas), a general surveillance on sea-going cargo and passenger ships of *all* nationalities calling at their ports or using their inland waterways in order to ensure that the requirements detailed in the Memorandum are met and that there are no conditions constituting health or safety hazards on board.

The six Conventions and Resolutions to be thus enforced are the Resolution concerning Standards on Merchant Ships adopted by the General Conference of the I.L.O. in its 62nd (Maritime) Session, 26 October 1976; the I.L.O. Merchant Shipping (Minimum Standards) Convention, 1976; Chapter 1, Regulation 19, of SOLAS 1960; SOLAS 1974; Article 21 of the 1966 Convention on Load Lines; Procedures and Guidelines for the Control of Ships set out in the IMCO MSC/219, 20 December 1976. Not all these Conventions are in force, and the Resolutions are not *per se* binding, which might be said to make them unenforceable on States not accepting them, unless they can be said to have been accepted into customary law, which remains doubtful. The parties to the Memorandum appear to consider that, as the standards are accepted by the parties and as they are not proposing to detain or prosecute substandard vessels, but merely to inspect them and report them to their flag States if inadequate under these instruments, and as recent Conventions such as MARPOL and the 1976 I.L.O. Minimum Standards Convention encourage and permit this form of limited application to vessels of non-party States, these actions are permissible under international law. It remains to be seen how rigorously the scheme will be applied, what effect it will have, and what protests it will provoke. Some objections to the regional schemes have already been vented in the *forum* of UNCLOS III.

3.2 EEC Actions[39]

The Hague Memorandum was concluded outwith the EEC; though it initially included five EEC members and the remaining two EEC coastal States (Eire and Italy) were invited to take part, two non-EEC members (Norway and Sweden) also participated. The EEC has, however, itself been engaged in intensive study of the problem of prevention of vessel-source oil pollution and has increased its efforts since the *Amoco Cadiz* disaster. The Commission has made a number of recommendations, including that EEC member States should simultaneously ratify all existing international agreements concerning oil pollution (making reservations in cases where enforcement provisions are considered inadequate); that they should accede to all existing regional agreements, including the Hague Memorandum; and that the eight coastal State members should extend their jurisdiction to a 12-mile limit for the territorial sea.[40] To improve detection and prosecution of offenders, a major problem in enforcement because of the constant movement of vessels through the ports and waters of many States, the EEC is considering, *inter alia*, negotiating through IMCO a right for the State detecting an offence to prosecute it, wherever it occurred, backed by inter-State exchanges of information (a kind of pollution Interpol on ICAO lines), and continual reporting-in by tanker captains of their position at sea to adjacent coastal States, the coastal State being given rights to participate in determination of the need for a pilot or tow.[41] Two draft Directives have been produced laying down minimum entry and exit requirements for Community ports for certain tankers (of more than 1,600 GRT).[42] The Commission also proposes a Council Decision making IMCO recommendations for ship inspection compulsory,[43] and compulsory pilotage for vessels navigating the North Sea and the Channel.[44] Adoption of EEC Directives and Decisions would have the advantage of making mandatory for member States IMCO Resolutions that are merely recommendatory.

The European Parliament has also interested itself in the problem of enforcement of many oil conventions, criticising present procedures, calling for Ship Traffic Control (STC) analogous to Air Traffic Control (ATC) and pointing out the dangers that 'ports of convenience' may develop if port State jurisdiction is not effectively exercised. A recent report recommended a regional agreement for control of tankers carrying dangerous substances; use of mandatory routes; a permanent co-operative EEC coastguard fleet to enforce all existing conventions; and measures to eliminate use of flags of convenience.[45]

Regional action is, however, still constrained by existing international law and change is inhibited by the general acceptance by the international community that it is in their interest to proceed in regulating such a valuable international commodity as shipping only by means of international agreement. The EEC and other regional bodies are therefore actively pursuing many of the above proposals at the international level as a Community in IMCO and especially at UNCLOS III. Let us finally, therefore, look at the progress made towards improving enforcement of pollution prevention measures in that *forum*.

4 THIRD UNITED NATIONS CONFERENCE ON THE LAW OF THE SEA: THE ICNT

The remaining weaknesses of the international legal system for enforcement of conventions to prevent oil pollution by vessels (apart from their poor ratification) lie in the still predominant position thereunder of the flag State. The main concern of reformist States at UNCLOS III has, therefore, been to redress this balance in favour of more port and coastal State jurisdiction. The subsequent negotiations on this topic have centred round construction of an equilibrium of their respective rights which will at one and the same time preserve maximum freedom of navigation and allow coastal States to protect their marine and littoral environments. UNCLOS is attended by up to 150 States and presents a unique opportunity (since IMCO has only 110 members and the I.L.O. 137) to establish a globally agreed regime for enforcement representing a balance of the interests of the international community as a whole.

It has held seven sessions to date and at the time of writing is in the process of an eighth. It is not yet negotiating on the basis of a treaty draft, but it has produced a series of informal texts for purposes of facilitating negotiation, the latest of which, the 1976 ICNT, is the current subject of debate by the Conference's Committee III, which is concerned with the Marine Environment, and the relevant Working Groups through which the Conference is at present conducting its negotiations. As no votes have been taken (as the Conference hopes to proceed by consensus) and no treaty adopted, it is impossible to evaluate at this stage the extent to which this Text, or the amendments being proposed to it, can be said to have entered into customary law. Many of the proposals have attracted considerable consensus, but not all and, in the opinion of the writer, it must at the present time be assumed that the enforcement proposals are still largely *de lege ferenda*. The relevant proposals are scattered throughout the Text's various chapters but read together, if adopted, would considerably improve the present regime. It is still intended that, as required by the UN General Assembly, the Conference should produce only one treaty relating to the problems of ocean space as a whole. The enforcement regime is one of the bargaining chips to be traded against other issues, such as the regime for the exploitation of the deep sea-bed, in the dealings for the final package; the proposals described below must therefore be read and related to each other with this qualification in mind.

4.1 Innocent Passage

Briefly, it is advocated that the territorial sea should extend to 12 nautical miles, that the right of innocent passage should be rendered non-innocent by the commission of any one of 12 specified acts, including 'any act of wilful and serious pollution, contrary to the Convention', and that passage must conform to international collision regulations and to coastal State laws concerning, *inter alia*, safety of navigation, regulation of traffic and preservation of its marine environment (including designation of sea lanes and traffic separation schemes); but not design, construction and manning or equipment of vessels unless such laws affect 'generally accepted' international rules or standards. Even warships (generally exempted from application of international rules) can be asked to leave the territorial sea if they do not observe the coastal State requirements for safe innocent passage. Somewhat similar requirements are laid down for regulation by the coastal State of 'transit' passage through international straits, which must conform to 'generally accepted' international regulations, including collision regulations.

4.2 High Seas

Freedom of navigation on the high seas would be preserved and, on it, vessels would be subject initially only to the jurisdiction of the flag State with which they have 'a genuine link', but the ICNT lists a number of specific duties to be assumed by the flag State to ensure safety at sea based, *inter alia*, on measures concerning construction, equipment and seaworthiness; manning, labour conditions and crew training, taking account of international Conventions. States with grounds for believing that these requirements have been violated can report the facts to the flag State, which must conduct inquiries following marine casualties.

4.3 Exclusive Economic Zone (EEZ)

The high seas would begin not, as at present, beyond the territorial sea, but beyond an EEZ extending 200 miles from the baselines of the territorial sea. In this EEZ the coastal State would exercise jurisdiction 'as provided in the Convention' for 'the purpose of preservation of the Marine Environment', though freedom of navigation and other rights of States (subject to these and other conventional restrictions) would be preserved.

4.4 Special Areas

It follows that very detailed specification is required elsewhere in the Text of the precise nature of the enforcement measures which the port and coastal State can take in relation to violations in their and other States' territorial seas and EEZs and on the High Seas. There is provision, first, for the designation of 'special areas', subject to appropriate international supervision and safeguards. In these areas, which are limited to the EEZ, a special mandatory system could be imposed by the coastal State because of the area's special vulnerability to pollution in the light of its oceanographic or ecological conditions, its use, or traffic, or the need for protection of its resources. The coastal State would regulate discharges and navigational practices, but *not* design, manning or construction of vessels. Ice-covered areas are dealt with separately and can be subject to an unfettered coastal State regime, thus meeting the Canadian interests in enforcement.

4.5 Flag State Rights

Enforcement in other areas of the sea is allocated in the Text by means of carefully graded port, coastal and flag State rights, the flag State's primary role being the more closely restricted in favour of the last two, the closer to the coast the offence occurs.

The flag State still retains the major role, but the Text requires States to establish international standards (through international conferences or organisations) for prevention of pollution from vessels, and for their enforcement. The flag State must ensure that the required certificates are carried, must inspect them and prohibit any vessel which does not comply with design, construction and manning requirements, from sailing. It must also investigate any violations it finds, or of which another State informs it, and take proceedings if the evidence is sufficient.

4.6 Port State Rights

The flag State's primacy is modified in favour of alternative, but not

exclusive, port State proceedings if a vessel which enters a port is found
by its authorities to have committed, on the high seas, violations against
international rules. If the violation has occurred in the territorial sea or
EEZ of another State, however, the port State can act only if that State,
or the flag State concerned, or a State threatened by the discharge, so
requests – unless the discharge in question threatens the waters of the
port State. The port State in these circumstances must investigate the
violation if another State so requests, but can itself take proceedings only
if the flag or coastal State concerned does *not* do so. Even then any
proceedings instituted must be transferred to those States if they so
request. Port States can detain unseaworthy vessels, which contravene
applicable international requirements for seaworthiness and threaten
damage to the marine environment; they can allow them to proceed to
the nearest repair yard but must let them leave immediately the repairs
have been effected.

4.7 Coastal State Rights

Coastal States would be permitted to take proceedings for any violation
of international rules which occurs in the EEZ or the territorial sea of
the State whose port the violator enters. Only if there are *clear grounds*
for believing that vessels in passage in a State's *territorial sea* have
violated national laws conforming to international rules for prevention
of pollution could the coastal State physically inspect the vessel, and
then it can only take proceedings against it if sufficient evidence is thus
revealed. If a similar offence occurs in the coastal State's *EEZ* it can only
require the suspected vessel to identify itself and give other relevant
information, unless the violation has resulted in 'a *substantial discharge*
into, and in *significant pollution* of, the marine environment', whereupon
the coastal State can, if the vessel does not co-operate in other ways,
physically inspect it. Only if a vessel commits in the EEZ a *'flagrant or
gross violation'* of international rules, or national ones affecting them,
resulting in discharges causing *'major damage* to the coastline or related
interests or resources of the coastal State' can it take proceedings, if the
evidence justifies this. The actual methods of inspection and other details
thereof, graded in terms of time and severity, are circumscribed by 'safeguards' detailed in the Text, but none of the terms emphasised is further
defined, which leaves coastal States with considerable discretion to determine
when their appropriate enforcement powers can be used.

The ICNT preserves the rights of coastal States to intervene following
strandings outside their jurisdiction which threaten their coasts, taking whatever measures are necessary to protect the coastline and related interests.

4.8 Additional Proposals[46]

Since, under present international law, port State jurisdiction, as defined above, does not exist, and there are no coastal State pollution enforcement rights in an EEZ or any other zone beyond the territorial sea, nor 'special areas' defined in ecological or other broad terms, and since even a coastal State's rights in its territorial sea are contentious, all the ICNT provisions mentioned would represent a considerable improvement on existing enforcement and enable such international Conventions as are in force to be effectively applied. The frequent use of the term 'generally accepted' to describe the international standards etc. which must be observed may even, if widely interpreted, mitigate the weaknesses of the present regime arising from poor ratification of many conventions. But many States, in the light of recent strandings giving rise to massive pollution, favour a still heavier weighting of the enforcement balance in favour of the coastal State and at the 7th Session of UNCLOS III further proposals to this end were made. These have been graded by the Chairman of Committee III into those which have achieved informal consensus; those which are not agreed but nearing consensus; and those which are not agreed.[47]

There is not space here to examine these proposals in detail. Suffice to say that in the first category are proposals that coastal States must adopt routing systems as necessary and that they should be notified of adjacent polluting or potentially polluting incidents. In the second category is a proposal that vessels committing flagrant or gross violations in the EEZ (as well, as at present, in the territorial sea) should be detainable by the coastal State, and also a proposal to 'legitimise' and extend the Hague Memorandum type of arrangements, by requiring that when a regional arrangement exists and when the ship in the territorial sea of a State applying it is bound for another State party, the flag State of the vessel concerned should indicate whether it complied with the port State entry requirements of the State to which it was originally sailing. As such an arrangement would require the approval of three States— the flag State, the coastal State whose waters the vessel traverses and the port State of original destination—this proposal remains contentious. It is doubtful how far the very delicate balance of flag, port and coastal State enforcement powers that has been constructed in the ICNT should, or could, now be disturbed without fatally disrupting it and dissipating the present painstakingly negotiated consensus.

4.9 Development in the UK

The slowness of entry into force of IMCO and ILO Conventions, the

small number of States that ratify many of them and the continuing
suspicion that some flag States still do not vigorously enforce such
Conventions as they have ratified has, following the *Eleni V*, *Christos
Bitas* and *Amoco Cadiz* strandings and the Sullom Voe discharge, led
many lobbies in the UK to propose further unilateral measures to
stiffen enforcement—such as that oil companies should write specific
instructions into the conditions of tanker charter requiring use of
approved routes, recognition of 'avoidance areas' where the volume
of traffic and ecological and other conditions do not warrant internationally negotiated special schemes;[48] and that port authorities might
include clauses requiring approved routes in their port entry contracts;[49]
that swingeing penalties of up to £50,000 should be imposed;[50] and
that the chartering oil companies should be liable for payment of
penalties incurred, and not the master or owner of the vessel, as at
present;[51] that unsafe tankers should be refused port entry if computer-
ised records establish their substandard condition and operations.[52] In
so far as such measures exceed those approved internationally as outlined
in this chapter, they could be recommended, but not enforced, by the
coastal State.

5 CONCLUSION

Governments of coastal States that are also maritime States are reluctant
to take such measures in view of the fact that they may be used against
their own ships and their crews in the ports of less scrupulous foreign
and far distant States; hence their willingness to accept the international
compromises which have made possible the existing qualifications on
flag State primacy in enforcement and the introduction of limited port
and coastal State jurisdiction. The slowness and the gradualness of the
moves towards these new forms of enforcement, and the detailed restrictions imposed on the new powers, are due to this split between the shipping
and environmental interests of many States. Tilting the balance further
away from the flag State will require governmental evaluation of factors
that are economic and political rather than legal. What is required now is
for the governments of the States concerned to face up to the economic
costs involved both for the shipowners and their enforcement agencies, and
to ratify and energetically to implement all the existing Conventions, which,
with the recent additions, provide the wherewithal for a vastly improved
regime for prevention of oil pollution of the sea; in other words, to bridge
the gap between entry into force and effective enforcement of rules and
recommendations.

NOTES

1. ACOPS (Advisory Committee on Oil Pollution of the Sea), *Annual Report 1977*, pp. 13 and 26.
2. For a description of the classical doctrine see C.J. Colombos, *The International Law of the Sea* (6th edn), pp. 62-7.
3. In force July 1958; 60 ratifications (as amended 1962 and 1969) by August 1978.
4. IMCO Doc., 15 September 1978, MISC (78) 7.E: *A Secretariat Study of Certain Legal Aspects of Intervention, Notification and Salvage in respect of incidents like the Amoco Cadiz.*
5. As in SOLAS 1974.
6. Tacit Amendment was included in the 1972 Convention on the International Regulations for Collisions at Sea; the 1973 Marine Pollution Convention and SOLAS 1974; a target date for entry into force was set for the MARPOL 1978 Protocol (June 1980); SOLAS 1974 and its Protocol (June 1979); an IMCO Resolution urging States to 'make every effort' to ratify SOLAS 1974 not later than June 1978 has *now* been complied with, even by the UK, by the passing of the 1979 Act.
7. Convention on the Territorial Sea and Contiguous Zone 1958; in force 10 September 1964; 45 ratifications by 1 January 1977.
8. Declaration on the Maritime Zone, Santiago, 18 August 1952, in *New Directions in the Law of the Sea*, Vol. 1, p. 231.
9. By February 1978, 68 States had asserted functional zones of up to 200 miles, mostly, apart from Oman which includes pollution control, for the purpose of controlling fisheries: list supplied to writer by Carroz and Savini (FAO).
10. Arctic Waters Pollution Prevention Act, 26 June 1970; *New Directions*, Vol. I, p. 199; US Statement on Canada's Proposed Legislation, ibid., p. 211. It has been suggested that Canada's 200-mile Fisheries Zone enables it to assert jurisdiction therein over pollution, as necessary, to preserve fisheries, and the Canada Shipping Act was being amended accordingly: Waldichuk, 'Canada's Proclaimed 200 Mile Fishery Limits and Environmental Implications', *Marine Pollution Bulletin*, 8(8), 1977, p. 175.
11. Art. 5.
12. R. Maybourn, 'Planning for Safety' (Paper No. 2), given at Safe Navigation Symposium, Washington, DC, USA, 17-18 January 1978.
13. Council of Europe, *Report on European Shipping Policy* Doc. 3662, 19 September 1975; Rapporteur Mr J. Prescott, p. 27.
14. Ibid. The Report lists States which have at times offered flags of convenience: Liberia, Panama, Honduras, Costa Rica, Lebanon, Cyprus, Somalia, Morocco, Singapore, San Marino, Haiti, Malta, Sierra.
15. But see Bates and Yost, 'Where Trends the Flow of Merchant Ships' in *The Emerging Regime of the Oceans* (ed. Gamble, Portecorvo), pp. 249-76. The 12 major maritime States on the basis of their Gross Registered Tonnage are Liberia, Japan, UK, Norway, USSR, Greece, USA, FRG, Italy, Panama, France, Sweden, roughly in that order: see also *Statistics Vital to Marine Safety*, DTI Report Series 3/1978 (March 1978); *Accidental Oil Pollution of the Sea*, Paper No. 8 (HMSO, 1976).
16. *Statistics Vital to Marine Safety*, DTI Report Series 3/1978 p. 7; extrapolated from Lloyd's Register of Shipping.
17. Lowe, 'The Right of Entry into Maritime Ports in International Law', 14 S.D.L.R. (1977) 597-622. Lowe denies that a right of entry exists as a general principle of international law, or as a custom thereof, but finds that a presumption favouring entry does exist. See also Lowe, 'The Enforcement of Marine Pollution Regulations', 12 S.D.L.R. (1975) 624-43.
18. Trade and Industry Sub-Committee (House of Commons Select Committee on Expenditure) (hereafter TISC) Session 1978-9, Vol. I, *Report on Measures to*

Prevent Collisions and Strandings of Noxious Cargo Carriers in Waters Around the UK; Ch. 5: 'The Human Factor', p. xxxvii, 'Oil Transportation Tankers and An Analysis of Marine Pollution and Safety Measures'; *OTA Report* (USA) 1975.

19. J.A. Sullivan, *Pollution of Harbours and Coastal Waters; Practical Enforcement of Pollution Standards in Harbours:* (b) *Milford Haven Case Study: Environmental Law* (ed. Novak), pp. 137-44.

20. Not in force; three ratifications (Kenya, Jordan, Tunisia) at January 1979.

21. By IMCO Conference on Tanker Safety and Pollution Prevention, February 1978; see note 24 below.

22. E.g. the recent Merchant Shipping Act 1979 was passed to enable the UK *inter alia* to ratify MARPOL.

23. International Chamber of Shipping Pollution Prevention Code (Oil Tankers) 1976.

24. International Conference on Tanker Safety and Pollution Prevention 1978 Protocol of 1978 Relating to SOLAS 1974, 7 March 1978 IMCO Doc. TSPP/CONF/ 10/Add. 1; Outcome of the Conference MEPC IX/5/Add. 1, 22 March 1978; Resolutions adopted by the Conference TSPP/CONF/12, 2 March 1978.

25. See also 'The Concept of Port State Jurisdiction', Report of British Branch Committee on the Law of the Sea, International Law Association, New Delhi Conference (1974) Proceedings.

26. See 'Winds of Change' (I.L.O. Publication); and Price, 'Tribute to Fifty Years Work of Seafarers' (I.L.O. Seafarers' Code, pub. M.N.A.O.A.).

27. International Convention on Civil Liability for Oil Pollution Damage; International Convention for Establishment of an International Fund for Oil Pollution Damage.

28. International Convention on Intervention on the High Seas in Cases of Oil Pollution Casualties, in force 1975; extended by Protocol to MARPOL to pollution by 'substances.other than oil'; not in force.

29. 'Ship's Captain Charged', *Scotsman*, 10 January 1979, p. 5; see also 'Chemical Fingerprints Lead to Oil Spill Arrest', *Int. Herald Tribune*, 10 November 1975, p. 1.

30. Cf. Tables 7 and 8, p. 44 in *DTI Statistics etc.*, note 16, showing a drop in contravention of Dover Strait Traffic Separation Schemes following the negotiation of the 1972 Collision Regulations, and the increased detection following improved radar surveillance in the UK; see also 'Channel Rogues in Decline': John Smith, UK Trade Secretary, reported that the average of 29 'rogue' tankers a day had now dropped to 16 following the entry into force in July 1977 of the Collision Convention: *Press & Journal*, 17 March 1978, p. 6.

31. Final Act, SOLAS 1974, ratified by the UK 1977.

32. Entered into force 21 July 1968.

33. I.L.O. Convention No. 76 (1946); revised by No. 93 (1949) and No. 109 (1958).

34. I.L.O. Convention No. 75 (1946); revised by No. 92 (1949) and No. 133 (1970).

35. I.L.O. Convention No. 145 (1976).

36. I.L.O. Convention No. 147 (1976).

37. I.L.O. Report V (i) and (ii); see also I.L.O. 62nd (Maritime) Session, Geneva 1976, *Record of Proceedings*, p. 230; Resolution on Sub-Standard Vessels, particularly those registered under Flags of Convenience.

38. Memorandum of Understanding between Certain Maritime Authorities on the Maintenance of Standards on Merchant Ships, signed at the Hague, 2 March 1978. TISC Evidence TT19 (COT Memo) Vol. III: *Europe*, 4 October 1978, No. 2531 (n.s.); the EEC Commission proposed that all nine EEC members should sign it.

39. A 'Background Report', 28 June 1978, ISEC/B47/78, points out the weaknesses of existing Conventions and recommendations and the difficulties of enforcement, but none the less opposes extension of 200-mile pollution jurisdiction, and supports regional 'practical' action.

40. *Europe*, 5 July 1978; and 'Marine Pollution Arising From the Carriage of

Oil (*Amoco Cadiz*); Commission Consultation Communication to the Council, COM (78/184 final, 28 April 1978).

41. Commission proposal for a Council Decision 'Rendering Mandatory the procedures for ship inspection forming the subject of Resolutions of IMCO', COM (78/580 final, 7 November 1978).

42. COM (78/586); OJ; No. C 284/5, 28 November 1978.

43. COM (78/587).

44. COM (78/580 final, 7 November 1978).

45. European Parliament (EP) Working Documents 147/78, 31 May 1978 (*Report on Commission proposals on marine pollution arising from . . . 'Amoco Cadiz'*); Doc. 555/78, 13 January 1979 (*Report on 1. the best means of preventing accidents to ships and consequential marine and coastal pollution; 2. shipping regulations*); Doc. 556/78, 16 January 1979 (*Report on Commission proposal for rendering mandatory the procedures for ship inspection forming the subject of resolutions of IMCO*). The reports recommend effective enforcement of all international Conventions dealing with all aspects of safety and conditions of work at sea; adherence by all member States to the Hague Memorandum; and effective port State jurisdiction.

46. These arose mainly during the Seventh Session of UNCLOS following the *Amoco Cadiz* disaster.

47. Reports of the Committees and Negotiating Groups on Negotiations at the Seventh Session contained in a single document both for the purpose of record and for the convenience of delegations: UNCLOS III, 7th Session, 19 May 1978, pp. 79-98; Report of Professor Yankov, Chairman, Committee III; UN Information Sheet BR/78/38, 29 September 1978.

48. *Press & Journal*, 16 March 1979, p. 6.

49. Letter to *Scotsman* from Dr J. Godfrey, 5 March 1979, p. 10.

50. *Scotsman*, 20 March 1979, p. 5. This was proposed as an amendment to the UK Merchant Shipping Bill.

51. Memorandum by ACOPS to TISC, Vol. III, Report, TT58.

52. *Scotsman*, 22 March 1979, p. 10. The Scottish Forth Ports Authority have turned away four unsafe ships using computer information, though the Chief Harbour Master states that 'tankers would not be turned away purely on their record, they would merely be checked more carefully'. EEC Council Resolution, 26 June 1978, recommends a study of the availability of such data in member States: OJ C. 162 Vol. 21, 8 July 1978.

4 THE ROLE OF THE OIL MAJORS AND OTHERS AS OIL TRANSPORTERS*

Gorden L. Becker

This chapter focuses mainly on the efforts of oil and tanker companies to provide compensation to those who have sustained pollution damage as a result of the escape or discharge of oil from tankers. It also discusses various steps these companies have taken to prevent such damage from occurring.

1 BACKGROUND

First, some background information on the scope of the problem. As pointed out by Yoshio Sasamura, Director of the Marine Environment Division of the Inter-Governmental Maritime Consultative Organisation (IMCO), principal maritime arm of the UN, in his booklet, *Environmental Impact of The Transportation of Oil*, during a twenty-year period beginning in 1954 there was a tremendous growth in the quantities of oil carried by sea, and a corresponding expansion of tanker tonnage to accommodate this increase. Between the mid-1950s and the mid-1970s the amount of oil carried by sea increased nearly seven times—from 250 million tons a year to 1 billion 507 million tons a year. This increase took place at the rate of about 10 per cent per year. At the same time, the world's tanker fleet, consisting of some 3,450 tankers of about 37 million deadweight tons, expanded to some 7,024 tankers of about 281,596,987 tons dwt. An interesting feature of this expansion was the replacement of many so-called T-2 tankers (approximately 16,000 tons dwt each) by tankers of 200,000 tons dwt or more. Indeed, these larger carriers (Very Large Crude Carriers or VLCCs) came to account for almost 40 per cent of the total tanker tonnage. The trend towards larger vessels resulted in part from the closing of the Suez Canal and the fact that large tankers cost less per ton of cargo capacity to construct and operate than small tankers. Since 1974, of course, there has been a lay-up of about 100 million tons dwt as a result of both over-building and a decline in oil demands. The increase in both the size of the world's

*The views expressed in this chapter are not necessarily the views of the author's employer. The author gratefully acknowledges the helpful suggestions made by his friends and colleagues, C.J. Carven, Susanna Opper and Richard Bavister, all members of Exxon's Logistics Department.

tanker fleet and the amount of oil carried has accentuated the problem
of marine oil pollution. Dramatic incidents involving pollution, like the
Amoco Cadiz, the *Christos Bitas*, the *Andros Patria*, the *Oregon Standard/
Arizona Standard*, the *Metula*, the *Torrey Canyon*, the *Urguiola*, the
Juliana, the *Arrow* and the *Argo Merchant* are well known. Less well
known is the fact that even more significant amounts of pollution have
resulted from operational activities. This pollution is from tankers that
have disposed of oily wastes at sea as a result of routine tank-washing.
At any rate, marine transportation, through casualties or operations,
has accounted for about one-third of all the oil put into the sea —
about 6.1 million metric tons each year. The rest is the result of
such activities as off-shore production, coastal oil refineries, natural
seeps, terminal operations and the like.[1]

The growth in the number of tankers, the growth in the size of tankers
and the impact of oil pollution has led to international, national and
industry efforts to deal with pollution. International Conventions,
generally under the aegis of IMCO, have attempted to cope with both
operational and accidental oil pollution from tankers and the compensation of victims of oil pollution damage. These Conventions (some not
yet ratified) include:

   (i)   the 1954 Pollution Convention and its Amendments, aimed
        primarily at restricting the amount of oily waste water that
        can be discharged and at proscribing any discharge in certain
        areas;
   (ii) the 1973 IMCO Pollution Convention (MARPOL), designed
        further to limit and prohibit discharges, to promote ship
        design and equipment that would lessen pollution from marine
        casualties, and to provide a means of monitoring discharges;
   (iii) the 1969 Intervention Convention and its Protocol permitting
        coastal States to take affirmative action against pollution or
        the threat of pollution originating on the high seas;
   (iv) the 1972 Convention amending the Collision Rules;
   (v)   the 1966 International Convention on Loadlines;
   (vi) the 1969 Civil Liability Convention (CLC) and the 1971 Fund
        Convention, which sought to provide recompense to the victims
        of pollution damage;
   (vii) the 1978 Protocols to the Safety of Life at Sea Convention and
        MARPOL, which tightened construction standards, prescribed
        crude oil tank-washing and segregated ballast tanks on certain
        tankers, required increased use of inert gas systems to prevent

explosions, and promoted improved methods for inspecting ships. For the first time some of these new standards were made applicable to existing ships as well as new ships.

In addition, the IMCO-sponsored Conference of July 1978 produced a Convention on Standards of Training, Certification and Watchkeeping for Seafarers, the first time that the international community has laid down basic regulations and recommendations on these subjects. Regional groups also have adopted Conventions and taken other action to combat pollution.[2] The Third UN Law of the Sea Conference has produced an Informal Composite Negotiating Text and related supplementary Reports which include various provisions on pollution and protection of the environment. For example, the provisions of Part XII of the Text dealing with vessel source pollution, if they become part of an international Convention, would go a long way towards increasing port and coastal State jurisdiction to take appropriate action against ships polluting or threatening to pollute the territorial sea or economic zone of States or failing to meet applicable standards. National governments also have taken unilateral action to minimise tanker accidents and the pollution resulting from maritime casualties. One of the best-known examples is the US Coast Guard's establishment of a system of vessel boarding inspection in American ports which was designed to prevent violation of prescribed standards. The US also passed legislation increasing the liability of tanker operators for clean-up costs incurred by the Federal Government and authorising the Coast Guard to implement international Conventions in advance of ratification and otherwise to strengthen rules applicable to tankers.[3]

2 TOVALOP

With this introduction to the pollution problem and some of the international and national efforts to cope with it, we move on to the action taken by oil and tanker companies in this area. Our principal attention will be devoted to TOVALOP and CRISTAL, tanker and oil industry regimes respectively, designed to compensate victims of pollution damage. After examining these agreements, we shall consider briefly certain other initiatives taken by these industries to prevent and abate pollution. TOVALOP is an acronym for Tanker Owners' Voluntary Agreement Concerning Liability for Oil Pollution. It came into effect in October 1969, at which time owners of some 50 per cent of the tanker tonnage of the free world had become parties to the agreement. TOVALOP was the brain-child of certain tanker-owners, many of them

also oil companies or oil company affiliates, who resolved to fill
voluntarily at least part of the gap that existed in traditional maritime
law regarding relief for oil pollution damage. TOVALOP in its original
form involved a commitment by each tanker-owner party, in the case of
a discharge of oil from one of his tankers, either to clean up the spill or
to reimburse a government for its reasonable costs in cleaning up the
spill. In short, the basic obligation of a party to TOVALOP was to clean
up or to pay – the payee being a government. But there were other
important provisions also. Of particular importance was a requirement
that each TOVALOP party carry sufficient insurance to permit fulfilment of its obligations. This insurance would also reimburse the tanker-
owner for his own clean-up expenses. There were certain limits on the
extent of a party's duty to pay, specifically $100 per gross registered ton
or $10 million, whichever was less.[4] Under the 1969 version of
TOVALOP, a party's obligation depended upon negligence of his tanker.
But despite the fact that negligence was an essential ingredient of a
party's liability, TOVALOP provided that it was *presumed* and hence
the tanker-owner had the burden of proving freedom from negligence,
an almost impossible task.[5] Various amendments and clarifications
have been made in the 1969 version of TOVALOP from time to time.
One of these was an amendment which made TOVALOP's provisions
applicable not only to a discharge of oil which caused pollution damage,
but also to the *threat* of such a discharge, even if no actual spill of oil
occurred. The TOVALOP party's basic obligation thus became a
responsibility to take reasonable measures to clean up a spill, to remove
the threat of a spill, or to compensate a government for its reasonable
costs for so doing.[6]

3 AMENDMENTS TO TOVALOP

On 1 June 1978 sweeping amendments to TOVALOP were put in
place which in many respects mirrored the provisions of the CLC which
came into effect in 1975 – six years after the inception of the original
TOVALOP. Some 38 states have ratified the CLC. The parties to
TOVALOP noted in the Preamble to the revision that although the
CLC had remedied many deficiencies in traditional maritime law for
which TOVALOP had offered substantial relief, there were major areas
in the world where the CLC did not apply and to which its benefits and
protection might not be extended for some time. Accordingly, pending
the widespread application of the CLC, the TOVALOP parties decided
to amend TOVALOP to provide benefits and protection generally
comparable with those available under the CLC in areas where the CLC

coverage was not yet available, and certain other benefits (discussed later). With the 1978 revision of TOVALOP and the coming into force of the CLC, there is now a reasonably uniform regime for tanker liability oil pollution damage world-wide.

Let us now look at the principal provisions of TOVALOP as revised. While making this examination, let us also note areas of difference between TOVALOP and the CLC. Each party to TOVALOP undertakes to compensate 'persons' (governments, individuals, or other entities) who sustain pollution damage as a result of the discharge of oil from a tanker owned or bareboat chartered by him.[7] Pollution damage is defined in almost the same terms as in the CLC, i.e. damage by contamination which occurs on the territory including the territorial sea of a State, including the cost of preventive measures to minimise or prevent such damage, and further loss or damage caused by such measures; remote and speculative damage is expressly excluded.[8] The liability assumed is strict, and the defences to it are limited — act of war, act of God, act or omission of third person with intent to cause damage, negligence of a government in failing properly to maintain navigation aids.[9] Negligence of the person who has been damaged will diminish or eliminate liability.[10] There is no liability whatsoever under the revised TOVALOP when CLC applies to pollution damage resulting from an incident.[11] TOVALOP further provides that each tanker-owner party will compensate persons who take reasonable measures to remove the threat of pollution after an incident occurs.[12] This coverage of 'pure threats', i.e. this commitment to compensate costs incurred in removing such threats, is not addressed by the CLC; the CLC relates solely to pollution damage resulting from an actual spill and to preventive measures taken to abate the results of a spill.[13] These tanker-owner obligations obtain whether or not the tanker was loaded — in contrast to the CLC obligations which cover only loaded tankers.[14] A tanker is defined in TOVALOP as any 'sea-going vessel and any sea-borne craft of any type whatsoever, designed and constructed for carrying oil in bulk as cargo, whether or not it is actually so carrying oil'. The limits of a TOVALOP party's liability in respect of a particular incident are $160 per limitation ton (about $147.00 per G.R.T.) or $16.8 million, whichever is less. These limits are roughly comparable to CLC limits. However, unlike the CLC, there is no contracting out of, or escape from, TOVALOP limits in the case of actual fault or privity of the tanker-owner — for example from the owner's putting the vessel in charge of an officer known to be incompetent.[15] The cost of measures taken by the tanker-owner himself to deal with a spill or a threat of a spill may be offset by him against the

limits of the liability which he has assumed. Thus, in a sense, the
TOVALOP owner can be claimant against himself, just as under the
CLC, and if the total claims against his limits of liability exceed such
limits, those claims – again after the CLC pattern – are prorated.[16]
Unlike the CLC, TOVALOP contains an express commitment by each
party to exercise his best efforts to take such preventive and threat
removal measures as are practicable and appropriate in the case of an
incident involving one of his tankers. In this respect the revised
TOVALOP reflects the same philosophy as the 1969 version of
TOVALOP – it strongly encourages the TOVALOP owner himself to
clean up and to avoid getting into disputes over claims.[17] Notice of
claim must be given within one year of the incident, and arbitration of
claims is subject to a two-year 'statute of limitations'.[18] Provision is
made for arbitration of claims under the Rules of the International
Chamber of Commerce.[19] Various definitions in TOVALOP generally
track or come close to those in the CLC. These include 'person', 'pollution damage', 'preventive measures' and 'incident'.[20] 'Oil', under
TOVALOP, means persistent hydrocarbon oil – crude, fuel oil, heavy
diesel oil and lubricating oil; TOVALOP is the same in this respect as
the CLC except that, unlike the CLC, it does not include whale oil.[21]
TOVALOP's definition of 'owner' means either the registered owner
or title-holder of the vessel or the bareboat charterer of the vessel; the
CLC is somewhat similar, but does not include a bareboat charterer.[22]
Each party to TOVALOP undertakes to apply the Agreement to *all* of
his tankers.[23] He further agrees to maintain financial capability to
meet just claims.[24] Normally he obtains liability insurance through a
'Protection and Indemnity Club', i.e. a mutual insurance association.
Obviously TOVALOP has the support of these underwriters.[25] As in its
original form, TOVALOP as revised is administered by a 'Federation'
(the International Tanker Owners' Pollution Federation Ltd) which
represents all of its parties and which has authority to lay down such
reasonable interpretive and other rules as are necessary for the proper
administration of the Agreement.[26] The Federation also performs
an additional important function – it provides the expertise of its technical staff to governments, underwriters and others in avoiding and
abating the deleterious pollution effects that can result from a marine
casualty. This staff has lent a hand in the *Amoco Cadiz, Andros Patria,
Christos Bitas, Eleni V, Betelgeuse,* the *Venoil-Venpet* and other incidents.

TOVALOP will remain in effect at least until 1 June 1981 and may, by
its terms, continue after that date subject to termination or withdrawal
by individual parties upon certain contingencies.[27] TOVALOP is

governed by English law.[28] It is difficult to make any accurate estimate
of the number of instances where payments have been made as a result
of TOVALOP (1969 or 1978 version), or the amount of any such payments, since the relevant data is not public information. It is reasonable
to assume, however, that a fairly large number of payments have been
made, including some which involve a number of well known incidents —
such as the *Arrow*, the *Oregon Standard/Arizona Standard*, the *Wafra*,
and more recently the *Imperial Sarnia* and the *Argo Merchant*. One
commentator credits it with helping dispose of 'hundreds of claims'.[29]
Under the original TOVALOP, many tanker charter parties included a
so-called 'TOVALOP Clean up Clause' which authorised the oil company
charterer to clean up at the owner's expense, at least when the owner
himself did not act promptly. It is reasonable to assume that the 1978
revisions to TOVALOP will prompt amendments to this clause and
that it will be tailored to the cut of TOVALOP and CRISTAL, as revised.
Although one can only guess as to what will be in a typical TOVALOP
clause prepared after the date of the revisions (and there will probably
be a wide variety of such clauses), one can certainly expect that the
charterer who properly cleans up or takes threat removal measures on
the tanker-owner's behalf will at least be reimbursed by the owner to the
extent permissible under the revisions. The question had been raised as
to whether the 1969 version of TOVALOP had any value in jurisdictions
such as the United States or Canada, which have their own particular
brand of pollution liability legislation. The answer given earlier probably
is still valid, that the existence of the revised TOVALOP gives such a
government an option as to which regime it may choose to utilise and
may suggest to some governments that TOVALOP is the preferable
means of relief.[30] In this connection, it will be recalled that some
governments, for example South Africa, the UK and Japan, despite their
own legislation, were disposed on occasion to rely on the 1969 version
of TOVALOP.[31] Even States which have ratified the CLC and which
have passed legislation to implement it may still find occasions where
TOVALOP will be helpful. For example, in cases where the vessel which
causes the spill flies a flag of a State other than one which has ratified
the CLC and the vessel is not otherwise covered by CLC; also in 'pure
threat' situations.

4 TOVALOP AND CLC COMPARED

We have already mentioned differences between TOVALOP and the
CLC, but it may be helpful to summarise the most important of these:

(i)   TOVALOP applies to *both* loaded tankers *and* tankers in ballast, unlike the CLC which covers only loaded tankers;

(ii)   TOVALOP, unlike the CLC, covers bareboat charterers as well as tanker-owners;

(iii) TOVALOP applies to the *threat* of a discharge as well as discharge itself;

(iv)   TOVALOP has a shorter 'statute of limitations' than the CLC, and claims that cannot be settled satisfactorily go to arbitration instead of to the courts as under the CLC;

(v)   TOVALOP does not permit an escape from the established limits in the case of actual fault and privity of the owner, as does the CLC;

(vi)   TOVALOP does not apply to whale oil;

(vii) TOVALOP, unlike the CLC, is under the aegis of an administrative body, the 'Federation', made up of the TOVALOP parties, and the Federation is charged with administering and interpreting the Agreement. This provides an encouragement for uniformity and a means of helping in the disposition of claims in difficult cases.

(viii) the Federation has developed in its staff a group of experts in the treatment and abatement of pollution who have been available in the case of large spills to render advice and assistance.

5 CRISTAL

We now move on to an examination of CRISTAL. CRISTAL is an acronym for Contract Regarding an Interim Supplement to Tanker Liability for Oil Pollution. It came into effect in April 1971, some seven years prior to the effective date of the International Convention on the Establishment of an International Fund for Compensation for Oil Pollution Damage. The parties to CRISTAL include oil companies shipping over 90 per cent of the world's total shipments of crude and fuel oil by sea. CRISTAL, in its original form, consisted of a Contract between its oil company parties and the Oil Companies' Institute for Marine Pollution Compensation Limited, a Bermuda association of which all oil company parties are shareholders. The Institute undertook to reimburse certain oil clean-up expenses incurred by tanker-owners, and also to compensate persons sustaining pollution damage resulting from tanker oil spills who would otherwise receive inadequate compensation. The oil company parties agreed to help the Institute in funds needed to make these payments. CRISTAL required, as a condition of liability, that the cargo be owned by one of its oil company parties, and that the tanker from which the oil escaped be owned by a party to TOVALOP. Further, for liability to arise, the escape of oil had to occur under circumstances where the tanker-owner either was or would have been liable for the resulting pollution damage under the CLC. CRISTAL imposed a per incident limit on the Institute's liability of $30 million. The Agreement made it clear, however, that any amount payable by the owner in discharge of his liability or from other sources were to be offset against this amount. CRISTAL, like TOVALOP, was modified so that the Agreement applied to the threat of a discharge of oil as well as to actual pollution damage resulting from a discharge.[32]

The June 1978 revision of CRISTAL made sweeping changes in the Agreement. Just as the 1978 revision of TOVALOP caused it to mirror to a large extent the CLC, the June 1978 revision of CRISTAL brought it fairly close in to the Fund Convention, although the latter did not come into force until late 1978. (The Convention now has been ratified by 17 States). The gist of these changes was the establishment of a system under which there was available from all sources, in respect of an incident involving pollution damage caused by a discharge of oil, or the incurring of costs to remove a threat of a discharge, compensation up to $36 million. This $36 million figure may be doubled if experience indicates that it is inadequate.[33] CRISTAL, as revised, continues to be a contract between its oil company parties and the Oil Companies' Institute for Marine Pollution Compensation Limited.[34] Under the Contract the Institute agrees to pay supplemental compensation to third persons (governments, private entities and other persons) who cannot collect in full under the CLC, TOVALOP or other regimes.[35] As in the earlier version of CRISTAL, the oil company parties respond to 'calls' from the Institute to put in it necessary funds to carry out its functions.[36] The Agreement envisages that the Institute will keep on hand the sum of approximately $5 million at all times although, of course, the system of calls can result in CRISTAL members having to respond for additional amounts.[37] CRISTAL definitions are generally similar to those in TOVALOP, except for the definition of 'tanker', which in CRISTAL excludes a loaded tanker. In this respect CRISTAL's definition is modelled on the CLC and the Fund Convention.[38] There are a number of prerequisites to the application of CRISTAL as revised to a discharge or threat incident. In the first place, the cargo must be 'owned' by an oil company party to CRISTAL and the tanker must be owned by a party to TOVALOP.[39] Second, none of the defences to liability available to an owner under CLC must exist. In other words, the basis of liability of the Institute is virtually identical with that of an owner under the CLC and under TOVALOP. Here CRISTAL varies somewhat from the Fund Convention, whose defences are limited to acts of war and the like, and in some cases at least in part to negligence of the damaged person.[40] Third, CRISTAL does not apply to pollution damage compensable under the Fund Convention.[41] A claimant, that is a 'person' (who, as we have noted in discussing revised TOVALOP, may be a government, an individual or any other entity), who sustains pollution damage or who incurs costs in taking threat removal measures, may obtain compensation under CRISTAL to supplement the amount of his recovery from other sources, provided that he has made reasonable efforts to so recover. His obvious sources of primary recovery are the tanker which discharged the oil (or which threatened a discharge) and whose liability is probably to be measured by the CLC, TOVALOP or some other regime, plus, in some cases, government funds created by a levy on oil companies.[42] Correspondingly, a tanker-owner who incurs preventive measures in an amount not wholly compensated under the CLC or TOVALOP limits may generally recover a portion of this 'pollution damage'.[43]

It was noted that one of the prerequisites of recovery under CRISTAL is that the cargo must have been 'owned' by an oil company party. Ownership is broadly defined and includes other types of ownership beyond the actual holding of title. Thus an oil company party may *deem* that a particular cargo was owned by him — by electing, before an incident occurs, to be considered the owner of a cargo transferred to a non-party, or if the oil is carried in a tanker owned or chartered by him or one of his affiliates. Furthermore, a party is considered to be owner when, prior to an incident, a non-party contracted to transfer title to the property to the party.[44]

CRISTAL's provisions supply important benefits for tanker-owners. We have already noted, for example, that preventive measures taken by a tanker-owner which exceed TOVALOP or the CLC limits may be reimbursable as pollution damage. But also, after the fashion of the Fund Convention, CRISTAL indemnifies a tanker-owner for a portion of his CLC liability and similarly indemnifies him for a portion of his TOVALOP liability or other liability between $120 per limitation ton or $10 million, whichever is less, and $160 per limitation ton or $16 million, whichever is less.[45] It also puts a tanker-owner in the same position he would normally be in under the CLC when a regime other than the CLC applies to the incident, this via special indemnification provisions.[46] There is a one-year notice of claim requirement.[47] CRISTAL is governed by English law and disputes are to be settled by the Courts of England.[48] CRISTAL will last until at least 1 June 1981 and may go on longer.[49] The Institute has the right to interpret the Agreement and issue Rules and directives.[50] One interesting example of the exercise of this power is shown in Rule 2.1 of the Institute which treats the Great Lakes as seas in determining if a tanker is sea-going. We have mentioned already some differences between CRISTAL and the Fund Convention. It seems appropriate, however, at this point to summarise these and mention certain other differences not previously examined.

CRISTAL liability is, as noted, subject to more exceptions than those available under the Fund Convention; perhaps the fact that CRISTAL is equipped to act faster than the elaborate Administration established in the Fund Convention makes up for this difference.[51] CRISTAL's period for notice of claim is shorter than that in the Fund Convention.[52] Further, the conditions under which CRISTAL indemnifies the ship-owner for a portion of his TOVALOP or CLC liability (commonly referred to as 'roll-back') are simpler than those provided under the Fund Convention.[53] Actual experience under CRISTAL has involved the pay-out of slightly over $6,100,000 and some 16 claims. It must be apparent, regardless of past experience, that the effects of inflation plus the possibility of valid claims for pollution damage in very large amounts indicate that CRISTAL, TOVALOP, CLC and Fund Convention limits should be substantially increased.[54] Also the existence of indemnity or roll-back provisions in CRISTAL and the Fund Convention is a 'built-in' complication and source of possible disputes between tanker-owners and those administering CRISTAL and the Fund Convention. It would surely be worth while to amend and delete these provisions.[55]

6 OIL INDUSTRY INITIATIVES

But, as must also be apparent, 'an ounce of prevention is worth a pound of cure'. TOVALOP and CRISTAL, like the CLC and the Fund Convention, deal primarily with situations which arise *after* a discharge has occurred or a threat of a discharge has developed and are designed to compensate those who have been damaged as a result. What kind of action, if any, has been taken by tanker-owners and oil companies to prevent pollution from occurring in the first place, to eliminate the threat of a discharge, and to abate pollution once a discharge has occurred? Both tanker companies and oil companies typically are members of various trade associations among whose purposes are the promotion of safer tanker operations. In the case of tanker-owners, various national associations which are concerned with these matters are members of the

International Chamber of Shipping (ICS), which recommends appropriate action on an international basis dealing with tanker safety and oil pollution. The Oil Companies International Marine Forum (OCIMF) is the leading organisation of which oil companies exclusively are members. OCIMF is concerned with tanker and terminal safety, protection of the environment and avoidance of pollution. Both ICS and OCIMF have consultative status with IMCO and have made a substantial input to that organisation in connection with such international Conventions as MARPOL and the Tanker Safety and Pollution Prevention Protocols of 1978.

In the pollution field, ICS has been responsible for the creation of such regimes as the ICS *Pollution Prevention Code* which is aimed at putting into practical effect the provisions of MARPOL, even though that Convention has not yet received sufficient ratifications or accessions to be in force. Other ICS activities include the publication of casualty reports which deal with tanker and navigational accidents as well as shipboard fires. These describe the relevant circumstances surrounding the incident, its possible causes, and lessons to be learned. The reports, issued in anonymous form, are circulated to masters and other officers in the hope that they will contribute to the avoidance of similar casualties. OCIMF (whose members account for about 85 per cent of oil transported world-wide by sea) has been responsible along with ICS and in co-operation with ICS for the production of such documents as the *International Oil Tanker Terminal Safety Guide* (recently amalgamated with the ICS *Tanker Safety Guide* and republished as *International Safety Guide for Oil Tankers and Terminals*) and other publications dealing with marine safety. These include booklets: *Malacca/Singapore Straits — Guide to Planned Passages for Draught Restricted Ships*; *Ship to Ship Transfer Guide (Petroleum)*; *Monitoring of Load-on-Top*; *Clean Seas Guide For Oil Tankers*; *Guidelines for Tankwashing with Crude Oil*; *Prevention of Oil Spillages through Cargo Pumproom Sea Valves*, etc. OCIMF itself has published booklets on *Tanker Safety and Pollution Prevention*, *Standards for Tanker Manifolds and Associated Equipment* and a series of booklets dealing with hoses used for Single Point Mooring. (An SPM is a facility which permits the moored ship to 'weathervane' around the mooring in response to forces of wind and sea prior to, during, and after cargo transfer operations.) OCIMF is also developing plans and strategies designed to strengthen the industry's capability to respond to marine casualties, including pollution incidents. OCIMF has been responsible for various research projects designed tó promote safety in shipping and thus, of course, to minimise risks of casualties which could cause pollution. A current example is a Hawser Test Project designed ultimately to assure safe mooring. Incidentally, the results of projects like this one will be made available to the public. In January 1978, OCIMF conducted a Safe Navigation Symposium at which experts drawn from government, industry and academic circles made available the fruits of their knowledge. IPIECA, the International Petroleum Industry Environmental Conservation Association, made up of oil companies and oil industry associations, works closely with UNEP (United Nations Environmental Program) and is concerned, *inter alia*, with the development of strengthening industry-government co-operation in dealing with oil spills, development of contingency plans, UNEP's 'Earthwatch Program' and the like.

The oil and tanker industries, sometimes acting individually and sometimes through various trade associations, have been responsible in large part for the development and improvement of systems dealing with Load-On-Top (LOT) and crude oil washing of tanks—commonly referred to as COW. Both of these techniques are designed to reduce the amount of oily mixtures which are discharged to the sea during routine cleaning of cargo tanks in preparation for clean ballast. LOT involves collecting the oily wastes and oily water from tank washings in a tanker's slop tank. The process uses the settling of the water and the rising to the top of the oil and the careful decanting of the water minus the oil on the run from discharge port back to the loading port. LOT affords tanker-owners a means of complying with the 1969 Amendments to the 1954 Convention, and its practice is recognised as a requirement in MARPOL. Properly practised, LOT is capable of recovering 99 per cent of the oil from oil-water emulsions resulting from tanker washings and dirty ballast.[56] COW involves the use by crude oil carriers of crude oil rather than water to wash tanks which have just carried cargo. The application of this system has minimised the need for water washing and therefore effected a further reduction in operational discharges of oily mixtures. COW has been 'blessed' by the 1978 Protocol to MARPOL. As Y. Sasamura noted in the booklet mentioned in the beginning of this chapter, 'crude oil washing will . . . make a significant contribution to pollution avoidance.' Although not aimed primarily at preventing pollution, the development of inert gas systems by oil companies and tanker-owners has been a prime contributor to the safety of ships discharging their cargoes. Inert gas pumped into the empty spaces of cargo tanks replaces air, thus eliminating explosive mixtures. The application of this system has been extended in the Tanker Safety and Pollution Protocols of 1978, i.e. from new tankers of 100,000 tons dwt or more as required in the 1974 Safety of Life at Sea Convention to all tankers of at least 20,000 tons dwt. Inerting is specifically required when tanks are crude washed.

As the cartoon strip character, Pogo, commented: 'We have met the enemy and he is us!' Tanker and oil company interests have recognised that while new and better equipment is needed to improve the safety of vessel operations and minimise pollution risks, the prime cause of marine casualties is human error. With this in mind, facilities have been developed by industry to improve the training of officers and crews. Prime examples are the facilities near Grenoble, France, and at Delft, Holland. Aboard miniature tankers at Grenoble, deck officers learn the safe handling of vessels and are put through a number of manoeuvres designed to sharpen their judgement. Delft involves a computer simulation of ship-handling situations. Like Grenoble, Delft is designed to sharpen the judgement of those taking the course and to familiarise them with the characteristics of different classes of ships. Tanker-owners and oil companies have taken steps to improve bridge organisation procedures aboard their ships. This has typically involved the preparation of detailed procedures whereby the exact duties of each officer are prescribed in writing rather than being left largely to the discretion of the master. Such procedures also typically involve a system in which voyages are pre-planned. Both the plan and its implementation are checked and cross-checked in an effort to eliminate one-man errors.

Oil and-tanker personnel have co-operated with governments to a large extent in the establishment of traffic separation lanes, for example in the Malacca Straits and the English Channel, and in procedures for safe harbour operation, for example in Rotterdam and Valdez. A number of oil companies, acting through OCIMF, have been studying means to improve the current regime applicable to salvage, a subject broached on a number of occasions in IMCO circles and now under study by various IMCO committees. These studies may recommend that salvors' duties include not only the salving of property (ship and cargo) but also the duty to abate pollution. In the future the salved interest may compensate salvors for actual expenses incurred in the salvage operations even when the operation does not succeed – a marked change from no cure, no pay. Industry leaders such as David Steel, Chairman of BP, have urged oil companies to take the initiative in developing various means for improving tanker safety and for avoiding and min-imising pollution. As of this writing, various projects looking to further these purposes are under consideration, including means to deal with problems involved as a result of the operation of ships which, because of their condition, may properly be considered 'substandard'. It seems fair to say that both the oil and tanker industries have accepted the challenge and recognised that further and continued efforts are necessary to improve tanker safety and drastically reduce and eliminate oil pollution damage.

NOTES

1. Booklet by Yoshio Sasamura, *Environmental Impact of the Transportation of Oil* (1977).

2. See *IMCO News*, No. 2 (1978) and *Lloyd's List*, 26 Jan. 1979, p. 7.

3. The Clean Water Act of 1977, P.L. 95-217, and the Port and Tanker Safety Act of 1978, P.L. 95-474.

4. See Clauses I, III, IV, V and VI of the 1969 version of TOVALOP. A detailed description of the pre-1978 versions of TOVALOP may be found in Becker, 'A Short Cruise on the Good Ships TOVALOP and CRISTAL', (1977) 8 J.M.L.C. No. 4.

5. Clause IV(B), 1969 version of TOVALOP.

6. See Clause IV(A) of TOVALOP as it existed just prior to June 1978.

7. Clause IV and Clause I(b).

8. Clause I(h).

9. Clause IV(B).

10. Clause IV(C).

11. Clause IV(B)(a).

12. Clause IV(A) and Clause I(k).

13. Art. I(6), II and III of CLC.

14. See TOVALOP, Clause I(a) and CLC, Art. I(1), defining 'tanker' and 'ship' respectively.

15. TOVALOP, Clause VII and CLC, Art. V(2).

16. TOVALOP, Clause VII(B) and (E). See CLC, Art. V(4) and (8).

17. Clause VI. See also 1969 version of TOVALOP, Clause V.

18. Clause VIII(C) and (D).

19. Clause VIII(I).

20. TOVALOP, Items (b), (h), (i) and (j) of Clause I and CLC, Items 2, 6, 7, and 8 of Art. I.

21. Clause I(g) and CLC, Art. I(5).

22. TOVALOP, Clause I(c) and (d) and CLC, Art. I(3).

23. Clause II(B)(1).

24. Clause II(B)(2).

25. See an excellent article by Bessemer Clark, 'The Future of TOVALOP', [1978] IL.M.C.L.Q. 79.

26. Clauses I(o), II, III and IX.

27. Clause III.

28. Clause XI.

29. See Fairplay, 'Who Pays and How Much', (1979) 10 J.M.L.C. 21; see also Becker, 'A Short Cruise'.

30. See 'A Short Cruise'.

31. Ibid.

32. A description of the 'original' CRISTAL and various changes made in it before 1978 may be found in 'A Short Cruise'.

33. CRISTAL, Clause IV(I).

34. See Preamble.

35. Clause IV(A), (E) and (F).

36. Clause VII.

37. Clause VII(B).

38. Clause I(A). Compare the CLC, Art. I(1) and Fund Convention, Art. 1(2).
39. Clause IV(B)(1) and (2).
40. Clause IV(C) and (D). Compare Fund Convention, Art. 4(1, 2, and 3).
41. Clause IV(A)(1).
42. Clause IV(E)(1) and (F)(1).
43. See Clause IV(E)(2) and F(2), and see also Clause I(G) and (H).
44. Clause V.
45. Clause IV(G).
46. Clause IV(H).
47. Clause VIII.
48. Clause XI.
49. Clause III.
50. Clause IX.
51. See Fund Convention, Arts. 16-34.
52. See Article 6 of Fund Convention and compare with CRISTAL, Clause VIII.
53. Compare CRISTAL, Clause IV(G) with Fund Convention, Art. 5.
54. Claims have been made against the Institute as a result of the *Amoco Cadiz* incident, but it is too early to assess the amount of actual damage involved.
55. For a further general discussion of TOVALOP, CRISTAL, the CLC and the Fund Convention, see pamphlet by Carven and Becker, dated February 1978, entitled *International Oil Spill Liability and Corporation Regimes* (published by Exxon Corporation).
56. *Environmental Science and Technology*, Vol. II, No. 12 (November 1977), p. 1046.

5 THE LESSONS OF THE EKOFISK *BRAVO* BLOW-OUT

Carl A. Fleischer

1 THE ACCIDENT

The Ekofisk petroleum field is situated in the Norwegian sector of the North Sea, fairly close to the median line between the Norwegian and the United Kingdom sectors. There are ten participants in the licence. Phillips Petroleum Company, Norway, with an interest of approximately 37 per cent is the operator.

On Friday 22 April 1977, at about 22.00 hours, an uncontrolled blow-out occurred on the production platform Ekofisk B – the *Bravo* platform – involving a spill of crude oil which was estimated to be something between 13,000 and 23,000 barrels per day. No such accident had earlier occurred in the Norwegian sector. However, the risk of serious damage to the marine ecosystem and to the coasts of the States bordering the North Sea was now obvious.[1] According to the figures given, the probable spill was between 2,800 and 3,000 tons per day; with a minimum figure as low as 1,700 and a maximum of 3,000. In addition there was a spill of gas, estimated to be between 1.4 and 1.6 m cubic metres per day. The figures given, which are based on estimates by the Norwegian Petroleum Directorate, coincide well with the figures based on calculations by Phillips as the operator.[2]

At the governmental and administrative level, the possibility of a blow-out had been considered in several documents. This includes the report to the Storting (the National Assembly) No. 25 for 1973/4, which was an endeavour to give a general survey and analysis of the impact of the petroleum industry and petroleum-based activities in regard to the Norwegian society. As is well known, these activities are of a fairly recent date on the Norwegian continental shelf, and there is no parallel experience concerning exploration for and exploitation of oil resources in the land territory of Norway. Consequently, the oil activities have been regarded as containing several unknown factors regarding the impact on the Norwegian economy as well as the possible adverse environmental effects, which might be the result of a major blow-out. The views put forward in St. meld No. 25 (1973/4) were, *inter alia*, as follows:

During drilling the greatest danger is the risk of explosive blowouts.

135

Normally these are caused by unexpected thrusts into an oil stratum
with such great pressure that it cannot be controlled.

The number of blowouts in comparison to the total number of
drill holes is small. A government sponsored commission of experts
in the United States concluded in 1969 that, statistically, there will
be one major blowout somewhere in the world every year. The risk
of a blowout occurring will depend to some extent on how comprehensive the geological exploration has been. However, there will
always be some uncertainty, as well as the danger of mechanical or
human error. Therefore, it is not possible completely to eliminate
every risk of a blowout on the Norwegian continental shelf.[3]

2 COMPANY AND GOVERNMENT RESPONSIBILITY

Under Norwegian law and the system applied in practice on the continental shelf, both the companies involved — and, in particular, the operator —
and the Government have important obligations to fulfil in order to
prevent accidents from happening and to combat the consequences if a
major oil spill should occur.

The basic principle is that the primary responsibility must be with
the operator. The Government may lay down regulations and it may
inspect installations and procedures from time to time, but it is the
operator who must perform the daily tasks of drilling and the building
up of installations and equipment. His performance must be within the
regulatory framework which is set up by the authorities, but that does
not relieve him of the obligation to exercise due caution in his activities
and the decisions thereon. Further, it is clear that the operator must
himself take steps to stop a blow-out when it has occurred and to
prevent, and if possible eliminate, the consequences. This also implies
that the operator must have contingency plans, and that he must acquire
and maintain the necessary equipment to take care of accidents and
dangerous situations. As is well known, the steps to bring a blow-out
under control may entail considerable expenditure, *inter alia*, when it
becomes necessary to drill a relief well and to contract a drilling rig and
personnel for that purpose.

Besides the primary obligations thus incumbent upon the operator,
the Government has the right — and the duty — to issue such directives as
may be found necessary, and also to approve steps taken by the operator,
to the extent that this is required by the regulations in force. Those
regulations also provide that the responsibility for the operation to
control the blow-out may be transferred to official authorities. They

may wholly, or partly, take over command of the operation from the responsible oil company, if the situation should make this necessary.[4]

The principle of primary company or operator responsibility concerns both the oilfield and such action as is taken at sea. If it becomes necessary to protect parts of the coastline or eventually to undertake clean-up operations after the oil has reached the shore, the matter is regulated by an act of 6 March 1970 on Protection against Oil Pollution Damage. Here, the point of departure is that the primary responsibility for protection of the shoreline lies with each single municipality (commune) affected. There is also an Oil Protection Council, which may direct operations at sea or involving damages at a larger scale. According to s. 2 of the Act, expenses may be recovered from the party responsible for the discharge of oil; but only to the extent that this follows from existing rules of civil liability.

Regarding civil liability, Norwegian law — as developed by the practice of the courts — is based on strict and unlimited liability for risk-creating activities. There seems to be no serious divergence of opinion as to the applicability of this jurisprudence to off-shore oil recovery and to pollution accidents caused thereby. A statute laying down the same principles has been under preparation for some time, awaiting the outcome of international negotiations and, in particular, the ratification of the 1976 London Convention.[5]

In addition to liability deriving from general principles of common or statutory law, undertakings by the licensees or by the operator may form a basis for obligations. In Norway, the partners in a production licence have to confirm the principle of strict and unlimited liability in writing, with particular regard to the claims of the State, including clean-up and other expenses.

3 THE OIL SPILL, ITS CONSEQUENCES AND THE SUCCESSFUL 'CAPPING' OF THE WELL

It may certainly be said that — in regard to the preparedness of the operator as well as that of the authorities — the *Bravo* accident occurred at an early and inopportune moment. The implementation of the applicable regulations had not progressed very far. The operator, Phillips Petroleum Company Norway, had not established its plan to meet the eventuality of an uncontrolled blow-out. A provisional draft of a contingency plan had been discussed between Phillips and the Petroleum Directorate, but no final plan existed.[6] The State Pollution Control Authority (Statens Forurensningstilsyn — SFT) had issued specific

requirements during the autumn of 1976 regarding mechanical equipment for oil collection on the continental shelf. Phillips had not taken
steps to carry out its obligations in respect of those regulations,
which, *inter alia*, involved the development of new types of equipment.
The companies had been given a time limit up to 1 August 1977 in order
to acquire such equipment, but when the accident occurred on the 22
April 1977, only prototypes of a new boom and a new oil skimmer had
been developed. Phillips did not have the organisation or the resources
which were necessary in order to carry out a large-scale oil-combating
operation.[7]

What has been said in regard to the company was also to some extent
applicable to the State and to the municipalities concerned. Work was
'in hand with building up the official oil prevention preparedness, following the guidelines which had been laid down by the Storting'.[8] The first
State depots containing oil prevention equipment had been established.
The intention was that the contingency plans of the State and the
municipalities should constitute a basis for measures at the coastal level
'which affords additional protection in the cases where the licence-
holders' combatting of oil spill at sea has not been able to prevent the
oil slick from drifting in towards the coast'.[9] But, as yet, there was no
collective national contingency plan. The programme was 'based upon
having established a satisfactory collective preparedness by the spring
of 1978'.[10]

However, if the Goddess of Chance had been somewhat unfavourable
in her choice of the moment when Norway was to have its first experience
of major oil pollution at sea, there were other factors which worked
to the advantage of the Government as well as that of the industry.
Because of the prevailing wind and current conditions in the Ekofisk
area and its surroundings, the oil kept itself at all times to the central
parts of the North Sea. In the first days, the oil stretched eastwards from
Bravo, and subsequently northwards. After about 12 May, it drifted
primarily southwards. The consistency of the oil underwent various
stages, and the oil amounts which could be observed on the surface were
gradually reduced. There were no reports of large amounts of oil from
Bravo after 13 June, though some was reported, in the form of oil lumps,
up to the end of July. There was no confirmed report of any oil from
Bravo having reached the coast of Norway or any other country.[11]

The Norwegian Institute of Continental Shelf Surveys (Instituttet for
Kontinentalsokkelundersøkelser – IKU) has in a report on the *Bravo*
blow-out estimated that about 40 per cent of the oil evaporated
immediately following the discharge. Further, the evaporation amount

rose to about 50 per cent during the course of 12 hours. After this the evaporation fell off. Fifteen days after the blow-out, the evaporated amount was estimated to be about 68 per cent. Possible further evaporation must have taken place slowly. The rapid evaporation, even before the oil fell on to the water, was primarily due to the fact that the oil had a high temperature (75°C), along with the oil being changed into small particles by the blow-out. The oil spread itself over the sea in the form of a thin brown membrane, up to 1 mm thick. At the outer edge of this membrane there were thick stripes of water-in-oil emulsion. The oil was also affected by large amounts of sea-water being sprayed against the oil-jet by a fire-fighting vessel (M/S *Seaway Falcon*).[12]

The Institute describes the conditions of the oil spill and the slick or slicks with reference to five stages: first, the oil spread itself out in the impact area in a fan formation from the platform, with a thin brown membrane, maximum 1 mm thick. The breadth of this oil membrane was about 100-200 metres and about 1 km in length. This oil membrane varied in extent with varying wind and wave effects (the membrane shrank in area with increased wind and sea). At the outer edge of this membrane there were thicker stripes of water-in-oil emulsion. Observations indicate that this emulsion was formed very quickly after the oil fell upon the water. Further, a thin oil film with a thickness between 0.01-0.001 mm was observed in large areas around the emulsion stripes. This film, known as 'blue shine', was very difficult to discover from ships, but could be observed from aircraft. At its height, the observed oil film (28 April) was contained within an area of 6,000 km^2. The fourth stage came after the blow-out had been stopped on 30 April and the wind died down. This resulted in large parts of the emulsion stripes being broken down. But, because of the strong effect of the sun, the oil underwent 'a hardening' process. It became considerably more viscous and sticky, and was more difficult to retrieve. The fifth stage was, after a relatively short time, when the emulsion stripes were broken up into small lumps about 0.2 to 2 cm in diameter. The Institute assumes that the amount of oil from the blow-out which might possibly have reached land in the course of July may only have amounted to about 10 tons.[13]

When the Main Rescue Centre Stavanger/Sola received a confirmed report that a blow-out had occurred on Ekofisk, about half an hour after the accident had taken place, key personnel from the involved administrations were collected and a provisional action command was established a few hours later. The authority of the action command was initially built upon oral decision by the Minister of the Environment, confirmed at a meeting with the Prime Minister and the Minister of Industry. Regulations governing the action command were established by Royal Decree on 26 April. The principal duty of the command was to follow and assess Phillips' plans and actions, draw up guidelines for the work, issue necessary instructions, orders and permissions to Phillips, and to co-ordinate and take more important decisions regarding the participation by official authorities.[14] The principle of the primary responsibility of the operator was adhered to.

Given a great amount of luck, a blow-out may cause such a fall in well pressure that the oil stream will peter out by itself. This did not happen in the *Bravo* case. Besides measures taken to contain the oil slick and to collect the oil by skimmers and otherwise, there are then two courses of action which present themselves, and which both have the objective of stopping the oil flow as such: there is the closing-off ('capping'), which may be carried out in a few days, and the drilling of one or more relief wells, which may take from one to several months. The first method proved to be feasible and successful in this case.

Already on 22 April, shortly after the blow-out had occurred, Phillips, as the operator, requested assistance from Red Adair Company Inc. (USA). This is a company which specialises in measures to control blow-outs. Two of the specialists from the Adair Company arrived at Stavanger in the afternoon on 23 April. A crane lighter, which was moved from the nearby Edda field to the area at the *Bravo* platform, was chosen as a working platform during the closing-off operation. The lighter was ready anchored in the immediate vicinity of the *Bravo* platform by the evening of 25 April. Phillips submitted its detailed plan in respect of the operation (including a contingency plan in case something might go wrong) for consideration by the action command during the course of that same day. After the plan had been studied by the Petroleum Directorate, it was approved by the action command on 26 April.

The flow was then closed on 30 April at 11.00 hours.

Obviously, the relatively fortunate outcome of events was largely due to this successful capping operation after only a few days. No relief well-drilling was necessary. In order not to lose time, should the attempt at a direct closing-off be unsuccessful, the action command on 23 April had ordered Phillips also to take steps to start the drilling of a relief well, simultaneously. On 24 April Phillips chartered a jack-up platform, which lay in Dutch waters, but this platform was not approved by Norwegian authorities (as a jack-up platform is dependent upon relatively good weather conditions). On 28 April the action command therefore decided that Phillips should take measures to bring a semi-submersible

platform to the field. The towing out of the platform then chartered by
Phillips commenced on 30 April; however, after the blow-out was
stopped on the same day, and the well was subsequently stabilised, the
action command issued permission for cancellation of the contracts
involving the hire of the rig. According to the action command, the
delay of about seven days from the decision taken (to launch also the
drilling of relief wells) until the commencement of the tow to the field
of a suitable platform was unsatisfactory; and could possibly have been
shortened if contingency plans had been worked out at an earlier stage.[15]

According to the head of the Red Adair Company Inc., Mr Paul 'Red'
Adair, he has 'capped in excess of 2,000 blow-outs, but he has never seen
so many people from the media covering the events'.[16] The blow-out
was, according to Mr Adair, not particularly difficult, and no catastrophe.
There were a number of favourable conditions in the case of the *Bravo*
blow-out. There was enough equipment; there was a barge; there were a
number of patrol boats. Besides, the well was not blowing any sand. The
Red Adair team only had two days of bad weather. However, he
claimed that more equipment and training was needed for workers in
the North Sea area, in order to be able to take the necessary steps
immediately if another blow-out should occur.[17]

4 THE RULES

The relevant rules are those which concern the prevention of accidents
as well as those which concern action to eliminate or mitigate the
consequences thereof; including the restoration of the environment to
its previous state, if possible. Also, the rules on civil liability are of
importance; in providing an incentive to avoid damaging consequences
as well as the funding of clean-up or similar action.

With regard to the *ex ante* situation – prior to the accident – important
regulations are contained in Royal Decree of 9 July 1976 relating to
Safe Practice for the Production etc. of Submarine Petroleum Resources.
This Decree has been issued by virtue of the Act of 21 June 1963,
concerning Exploration and Exploitation of Submarine Natural Resources,
s. 3. In particular, reference may be made to Chapter 11 (s. 92 to 110),
concerning 'production drilling'. According to s. 94, the mud system
shall be such that the well is kept under control at all times, and there
shall be blow-out preventers 'able to withstand any forseeable pressure
in the well' (s. 95). 'All necessary steps' shall be taken to prevent
explosions, blow-outs, pollution or other damage (s. 101).

This Decree also contains special regulations concerning the obligation
of the operator to prepare for the possibility of accidents, to train

personnel and to maintain equipment to be used in such eventualities. S. 39 has the following wording:

> The licensee shall maintain at all times a state of preparedness making it possible, in the event of an accident or dangerous situation, quickly to bring the situation under control and minimize the damage caused by such accident or dangerous situation.
>
> The licensee shall prepare an emergency plan for use in the event of accidents or dangerous situations. The plan shall cover the following main headings:
>
> a) Situations that have involved or may involve personal injury, serious illness, or loss of human life.
> b) Situations that have involved or may involve pollution.
> c) Situations which have or may put all or parts of such facilities as mentioned in Section 2 above out of service.
>
> The Ministry may decide that the emergency plan shall comprise other accidents and dangerous situations besides those mentioned above.[18]

Regarding the *ex post* situation — after a blow-out has occurred — it has been provided by s. 18 of the same Decree:

> In the event of an uncontrolled blow-out or other uncontrolled escape of petroleum, the licensee shall effect forthwith the necessary measures to bring the situation under control, to minimize the damage, and to carry out the work necessary to restore the environmental conditions as far as possible to the same state as before the escape occurred. Authority in respect of the collection and removal of petroleum spills is vested in the Ministry of the Environment.
>
> Spills shall be removed primarily by mechanical collection. Chemical dispersants may be used only as decided by the Ministry of the Environment.[19]

S. 46 empowers the authorities to order the drilling of relief wells at the licensee's expense.

As regards the aspect of civil liability, the relevant principles of Norwegian law have been mentioned above in connection with the relationship between company and State responsibilities.

5 LESSON ONE: THAT THE RISK OF A MAJOR BLOW-OUT EXISTS AND MUST BE TAKEN INTO ACCOUNT IN PRACTICAL PLANNING

From time to time one has encountered views at conferences, in the press, in political statements, etc., to the effect that the risk of a major blow-out in the North Sea was negligible. At any rate one has had to face an extreme uncertainty when the blow-out risk has been offered as a reason for a certain decision or conduct, and in particular for refusing the opening of new fields or for the incurring of expenses concerning extensive contingency plans. Was there really such an animal as the risk of a major blow-out, and could such a risk justify measures having important economic consequences for the companies and for society? Extra expenses for contingency plans put upon the operators would also partly have to be borne by society. This is, in particular, a result of the mechanism of taxation and deductibles. Further, increased expenses have an adverse effect on the calculations as to whether marginal fields should be developed for production; which, in turn, influences the energy supply of the society as well as its benefits from taxation and royalties. Against this background, it seems fairly natural that politicians, administrators, company representatives and lawyers alike must have asked themselves the crucial question: was the major blow-out possibility only a product of human imagination; a monster, like a griffin, against whom extensive precautionary measures were taken only as a result of lack of knowledge?

The *Bravo* blow-out goes a long way in answering those questions, even if not in the direction one could have wished. A major blow-out and massive pollution by oil are indeed factors which must be taken into reckoning when drilling for hydrocarbons in the North Sea, and even in areas further to the north in the future.

6 SECOND LESSON: THE POSSIBILITY OF ACCIDENTS HAVING FAR MORE SERIOUS CONSEQUENCES IS NOT TO BE RULED OUT

A number of factors served to minimise the damage caused by the *Bravo* incident. The well and the installation were intact. The oil did not catch fire. It was possible to cap the well and to stop the flow by using the *Bravo* platform and existing structures. Available specialists and equipment were called in, and the flow was stopped after only a few days. The oilfield is situated in the very middle of the North Sea, close to the median line. Prevailing winds and currents, and even a change in the direction of the wind, made the oil slick stay in the vicinity of the Ekofisk field, until it was eventually broken down. At least, it was no longer present in the form of an oil slick capable of causing large-scale ecological damage to the shoreline of a riparian country. Extensive use of chemicals was not made necessary. Only a relatively small amount was used, to reduce the risk of the oil catching fire,[20] which might have made the capping operation impossible. Further, weather conditions were reasonably good.

It is obvious that these factors may not be present in a possible future blow-out situation, be it in the Norwegian sector or elsewhere. In particular, it has been pointed out that operations in areas further to the north may imply greater hazards.

Further, the Ekofisk experience may have served to demonstrate that the existing mechanical equipment for containing the oil slick and for collection of oil was not satisfactorily adapted to the heavier sea conditions which may be expected in the North Sea. The development of more efficient types of booms and skimmers may be expected. But one may, if another major blow-out should occur in the present stage of technical development and under conditions less favourable then at the *Bravo*, be faced with the crucial choice between extensive contamination of the shoreline on the one hand and the large-scale use of chemical dispersants on the other, with possible adverse consequences for the marine ecology.

7 THIRD LESSON: THE HUMAN ERROR FACTOR

The official Commission of Inquiry, which was set up as a result of the accident, arrived at the following main conclusions:

- to a large extent the blow-out must be ascribed to human error
- certain technological weaknesses are revealed, but these were only of peripheral significance for the course of events
- the underlying cause of the accident was that the organisational and administrative systems on this occasion were inadequate in respect of the planning and management of the work, the directives for its performance, the formal routines for inspecting and reporting, detecting indications of error and effecting countermeasures.[21]

Later, those conclusions received general support in the National Assembly; cf. in particular the Report of the Storting Committee for Industry of 10 November 1978.[22]

More specifically, the Commission pointed to the following:

Thus the blow-out must be ascribed to a series of circumstances which to varying extents were direct or indirect contributory causes. It occurred as the result of a combination of

— undesirable practices as regards documentation for the installations (the tubing hanger), and as regards the marking of equipment (BMH plug) and instructions for its use (Baker DHSV in mud)
— weaknesses in the workover programme (did not prescribe the pressure testing of mechanical means of well control)
— undesirable organization of work on board the platform (for installation of BOP)
— failure to respect the established programme (omission of circulation and improvisations when it was learned that no back pressure valve could be fitted in the tubing hanger)
— critical situations were misconstrued (when last DHSV was fitted, when mud came up through the control line, and when mud came up through the tubing)
— weak management and supervision of the work (no leading personne — the drilling supervisor, the drilling engineer and the toolpusher — were on the decks where work was in progress during the whole of the most critical stage of the work, from 1800 hrs until the blow-out occurred)
— unreasonably long working hours for some personnel (wireline operator and drilling supervisor).[23]

8 FOURTH LESSON: MEASURES TO BE ADOPTED AT COMPANY LEVEL AND IN DOMESTIC LAW

Generally speaking, it may seem that the existing regulations which had been laid down in Norwegian law can be regarded as satisfactory. It was not due to lack of legal obligations that the operating company — and even public authorities — were not fully prepared and equipped to deal with a major blow-out, but because of the fact that the incident took place when they were still in the process of implementing the regulations. In particular, the required contingency plan had not been finalised.

It goes without saying that the human error factor can never be completely eliminated. An occurrence such as the *Bravo* will, however, usually assist the companies and the authorities in pinpointing such elements of existing operational routines where human error may have its more serious consequences, and where additional safeguards may be required. In this light, the *Bravo* accident may prove to have been a valuable practical experience contributing to a higher degree of safety in future operations.

In particular, reference may here be made to the more detailed findings of the Commission of Inquiry quoted above.[24]

The experience of the *Bravo* accident does not seem to justify any change of the basic principle of the operator's responsibility for the daily operations at the platforms and for the steps to be taken in case of accidents. This view has been endorsed by the conclusions of the Storting's Committee on Industry. In its conclusions, the Committee states strongly that the responsibility to take measures against the oil flow after an accident occurs must in the first instance be with the operator responsible for the activity which has led to the accident.[25]

According to a minority view in the Committee, the *Bravo* accident should result in a postponement of the plans for exploratory drilling north of the 62nd parallel.[26] This was, however, not accepted by the majority.

9 FIFTH LESSON: THE NEED FOR STRICT AND UNLIMITED LIABILITY

The *Bravo* incident also serves to reinforce the arguments advanced for strict and unlimited liability. Only such a system can give full cover for the victims and secure that the operator and the licensees will have the maximum interest in preventing damage, as well as the funding of clean-up expenses. As the incident demonstrates that a major blow-out may occur, and further that one can have no certainty as to the extent and the duration of the oil flow, it seems that any limitation on liability may have the consequence that sooner or later one will reach the time when damage is no longer attributable to the companies in question, when damage (including future clean-up) has reached the order of the limitation amount and damage in excess thereof will have to be borne by the victims or by the Government.

As mentioned above, the system of strict and unlimited liability is already provided for through general principles established in Norwegian jurisprudence.[27]

From time to time arguments have been advanced in favour of limited liability.[28] In the opinion of the present writer, those arguments are not tenable;[29] and this has been further proved by the *Bravo* incident

From a general point of view, it seems that provision should be made for compensation to the victim for *all damage* sustained as a result of the operations. The reasoning behind the principle of strict liability is that the damage caused by a given activity should be borne by the owner or operator as a cost of the activity, rather than by the unfortunate victim who may happen to suffer damage. The same general principle has, *inter alia* in the Council of Europe, found general acceptance as the so-called 'Polluter-Pays-Principle'. This principle may be said to apply whether or not damage is above a certain fixed limit.

To the extent that the rules of maritime law may be applicable, a limit on liability to a fixed amount may be laid down even under existing law. It does not seem probable, however, that a drilling rig in operation – still less a platform – could be regarded as a ship. In any case, the limitation following from such a viewpoint would only apply to the owner or operator of the rig, not to the licensee or owner of the oilfield.

The historical precedent of the limitation of shipowners' liability is not necessarily relevant to off-shore operations, where there is no similar traditional system of limited liability creating expectations of continuity. Here, there are reasons to apply the general considerations with regard to the question of who should carry the burden of the expenses of the industry.

The objection might be made, from the view of equity, that it may be too severe a burden upon the industry to make it strictly liable without any limitation. This argument has indeed a *prima facie* validity, as the imposing of a severe burden of liability is never equitable in itself. On the other hand, we must ask whether it would be more equitable to let the burden lie on the third party which happens to suffer such damage – quite frequently major damage – as the result of an oil spill.

A second reason for limited liability in this field is of a purely practical and political nature. It may be that the risk of unlimited liability will deter oil companies, or at least certain companies, from taking part in operations on the continental shelf.

Whether there is any substance in such objections to unlimited liability is a question that can only be answered on the basis of practical experience. It seems that the situation of the North Sea shelf is here particularly relevant. Oil companies have had to conform to the present Norwegian rules on strict liability, and to confirm the principle of strict and unlimited liability in their acceptance of the conditions for oil and gas concessions. Nevertheless, there has been no tendency for companies to refrain for this reason from operations on the Norwegian shelf or to withdraw therefrom. It seems a tenable contention that the rules on civil liability have little relevance for decisions as to whether or not a company shall take part in operations on a certain continental shelf and apply for a certain licence.

In this connection, it must be borne in mind that most countries, even if they do not practice strict liability, will have rules on vicarious liability.

Such liability will in practice be unlimited. With the exception of ship-owner liability there is usually no legal ground for limiting the liability of a master so that he will only be liable for part of the damage caused by one of his servants. No oil company has any guarantee that some employee or other may not be at fault in connection with an accident, or at any rate that a court may not find that this is the case. One may even say that there is a presumption of fault if something goes wrong, causing a major accident, as this should not have happened if everybody on the platform had done what he was supposed to do. If the risk of heavy costs as a result of unlimited liability was sufficient to cause a company to refrain from off-shore operations, or in other words if the leaders of oil companies slept that badly at night, one might have expected that they would have withdrawn from any operation anywhere in the world as long as there existed the principle of vicarious liability for faults committed by servants. Indeed, the difference between strict liability and liability based on fault may not be very great in regard to the number of cases leading to liability under the two different systems. For example, it has been said that Danish courts tend to establish liability by finding negligence in most cases where, if the case had come before a Norwegian court, the general principles on strict liability (which are not formally recognised in Danish law) would have been applied. Therefore, one cannot expect that a country with unlimited fault liability will be much more attractive to the oil industry than will a country with unlimited strict liability. This difference does not constitute an important factor in serious considerations by the industry as to whether or not a certain possible oilfield or natural gas source should be developed, and as to what amounts should be invested for this purpose.

The practical effect of a limitation is that damage caused by the operations in excess of the fixed limit will have to be borne by the victim or by the Government, while the company or companies involved would at the same time be able to take out the profit from the field. This would apply for a company having invoked the limitation, if an accident should indeed occur which resulted in losses exceeding the limitation figure. In cases where the aggregate losses come below that figure, the limitation has no practical effect whatsoever—liability might as well have been unlimited. Compensation is paid *in toto*. It may be said that it is inherent in the character of a limitation on liability—viewed from the *ex post facto* angle, after the accident has happened—that it will always and *per se* curtail the rights of the victims and enable the responsible party to take profits from his operations without paying the expenses involved in full.

10 INTERNATIONAL RULES ON CIVIL LIABILITY: THE 1976
LONDON CONVENTION ON CIVIL LIABILITY FOR OIL
POLLUTION DAMAGE RESULTING FROM EXPLORATION FOR
AND EXPLOITATION OF SEABED MINERAL RESOURCES

The 1976 London Convention was only opened for signature as late as
1 May 1977, and was not in force at the time of the *Bravo* incident.

It may, however, be of a certain interest to note the significance of
the Convention's provisions with regard to an accident of this character.

First, it seems that the accident serves to commend the steps taken
by the North Sea States in order to work out a suitable system of civil
liability with regard to pollution from off-shore operations.

Second, one may mention as important provisions the definitions
in Art. 1 of 'pollution damage' and of 'preventive measures' (which is
included in the wider term 'pollution damage'). It should be noted that
the definition of 'preventive measures' (and thereby 'pollution damage')
makes an express exception for 'well control measures and measures
taken to protect, repair or replace an installation' (Art. 1, paragraph
7). Such measures fall outside the scope of the Convention's liability.
This means, *inter alia*, that the cost of, for example, drilling a relief
well, cannot be deducted from the limitation fund, but must instead be
carried out by the operator at his own expense. Art. 1, paragraph 6 and
7 and Art. 3 must here be read in conjunction with Art. 6, paragraph
10: 'An operator who has taken preventive measures shall in respect of
those measures have the same rights against the fund as any other claimant.' Further, reference may here be made to the fact that Art. 1,
paragraph 6's definition of 'pollution damage' is limited to 'damage
outside the installation'.

The drilling of relief wells, the hiring of rigs and other equipment
for this purpose etc. are parts of the operator's or licensee's obligations,
irrespective of the Convention's provisions on civil liability as such. The
expenses incurred are not subject to any specific limitation and cannot
be taken from the limitation fund. In consequence, the preventive
action will not be hampered by lack of funds, by the need to decide on
a possible priority between preventive action at the well site and
compensation to property-owners and other victims at the shoreline,
fundings of their possible remedial action, etc. One does not have to
consider the necessity of governmental appropriations, and whether the
money which is available should be used for the one or the other purpose.

Third, one must take account of the fact that the 1976 Convention,
as a compromise, has been based on the principle of limited liability as
its point of departure. But the limitation amounts set out in Art. 6 must

here be read in conjunction with Art. 16. Para. 1 of this latter article reads as follows:

This Convention shall not prevent a State from providing for unlimited liability or a higher limit of liability than that currently applicable under Article 6 for pollution damage caused by installations for which it is the Controlling State and suffered in that State or in another State Party; provided however that in so doing it shall not discriminate on the basis of nationality. Such provision may be based on the principle of reciprocity.

In the opinion of the present writer, the experience of the *Bravo* incident clearly shows the advisability of this exception to the Convention limitation; exceptions which were also, at that time, proposed and supported by the Norwegian delegation at the 1975 and 1976 Diplomatic Conference in London.

If one had here instead yielded to the views advocated from certain quarters to the effect that Norway should have a system of limited liability in its domestic law, or to the effect that one should follow a system of limited liability if this was laid down in other countries like the United Kingdom, the Norwegian authorities in a case such as that of *Bravo* would have been forced to take their action with a view to the possible insufficiency of the funds available. It would have entailed considerations of priority between different purposes for governmental appropriations; instead of immediate and effective action with the overriding purpose to stop the blow-out and to protect the marine environment.

The aim of achieving uniformity between different national legislations cannot be regarded as the only relevant purpose, to be given priority before the protection of victims. The cost of uniformity may be that victims of pollution will be denied full compensation in the entire Convention area, because there is one single party which does not feel disposed towards accepting a sufficiently high level of obligations to be put upon the industry.

11 FURTHER INTERNATIONAL IMPLICATIONS

In the field of civil liability for off-shore operations, the adoption of the 1976 London Convention may be regarded as part of the implementation of Principle No. 22 of the 1972 Declaration of the UN Conference on the Human Environment:

States shall co-operate to develop further the international law
regarding liability and compensation for the victims of pollution
and other environmental damage caused by activities within the
jurisdiction or control of such States to areas beyond their jurisdiction.

The *Bravo* incident probably does not justify any amendments of
the limits for liability or for insurance (cf. Art. 9 of the Convention
on a Committee to be set up for this purpose, and which may recommend changes *inter alia* because of 'information concerning events
causing or likely to cause pollution damage'). However, the incident
serves to demonstrate the need for co-operation, and to ensure the
adequate protection of victims.

The continental shelf is not as such a 'shared resource'; or, rather, as
far as its 'exploration' and 'exploitation of its natural resources' are
concerned, the shelf is subject to the sovereign rights of the coastal
State.[30]

As the coastal State's rights are limited to those specific purposes —
if still 'sovereign' — it follows that the area may in other respects be
subject to the traditional principles of the high seas and thereby a *res
communis* (or, in modern terminology, a 'shared resource'). In particular,
the rights of the coastal State over the shelf do not affect the status of
the superjacent waters.[31] It follows further that off-shore petroleum
production which is carried out under the sole jurisdiction of the
coastal State, and which leads to a blow-out, may have direct effects on
the high seas fisheries of the area, to the detriment of other States whose
rights to the living resources are equal with that of the coastal State.
Further, the oil may drift towards the coasts of other States, damaging
resources which are subject to the sovereignty of those States.

The advent of the exclusive economic zone may, to some extent, have
changed this picture. Even if the rights over the shelf will not *per se*
influence the status of the superjacent waters, the living resources will
by now as a rule be subject to coastal State sovereign rights because of
the establishment of exclusive economic zones (or fishery zones) up to
200 nautical miles. But even with this development — which, at least
according to the views of the present writer, has now a basis in customary international law[32] — there may be fishery rights accorded to other
States within the zone. And the oil may, for example by drifting across
a median line, soon be found in areas subject to the sovereign rights of
States other than that responsible for the installation where the blow-out
has occurred.

Against this background, it seems clear that international law must have at least some fairly general principles obliging the coastal State to take due regard to the interests of others and, in particular, to prevent damage; in the *ex ante* situation, through the laying down of safe practices, as well as *ex post facto*, after a blow-out has occurred.

Such principles may be found not only in the law concerning the continental shelf as such – or, more generally, the law of the sea – but have their basis in more fundamental considerations of international law,[33] as indeed evidenced as early as in 1938 and 1941 in the *Trail Smelter Case*.[34]

Reference may further, *inter alia*, be made to Art. 5, paragraphs 1 and 7 of the Geneva Convention on Continental Shelf:

1. The exploration of the continental shelf and the exploitation of its natural resources must not result in any unjustifiable interference with navigation, fishing or the conservation of the living resources of the sea, nor result in any interference with fundamental oceanographic or other scientific research carried out with the intention of open publication.
7. The coastal State is obliged to undertake, in the safety zones, all appropriate measures for the protection of the living resources of the sea from harmful agents.[35]

In the 1972 Declaration of the UN Conference on the Human Environment Principles 7 and 21 have the following wording:

7. States shall take all possible steps to prevent pollution of the seas by substances that are liable to create hazards to human health, to harm living resources and marine life, to damage amenities or to interfere with other legitimate uses of the sea.
21. States have, in accordance with the Charter of the United Nations and the principles of international law, the sovereign right to exploit their own resources pursuant to their own environmental policies, and the responsibility to ensure that activities within their jurisdiction or control do not cause damage to the environment of other States or of areas beyond the limits of national jurisdiction.

Those general provisions are, however, probably too vague to shed any light on the conduct of the authorities and the companies in the *Bravo* case. Obviously, the 'all possible steps' in Principle No. 7 cannot be taken literally, as this would entail a total prohibition against all off-shore activities as long as there is no 100 per cent certainty that accidents and human error can never occur. 'Possible' must, further, be related to a certain test of reasonableness, in particular with regard to the expenses that may be incurred.

However, it would seem to follow from the above provisions that a State involved in exploration and exploitation of the resources of the continental shelf must review the facts and the rules, after having experienced an accident such as the *Bravo* blow-out, in order to arrive at safer practices in the future. The extensive investigations which were carried out by the Commission of Inquiry[36] may be regarded as part of the fulfilment of this environmental obligation. Further, some degree of international co-operation may be warranted in order to protect the different national territories as well as the high seas. Discussions had here, independently of the *Bravo* incident, been carried on continuously between the North Sea States, following a conference arranged in London in 1973 at the invitation of the United Kingdom Government.[37]

NOTES

1. According to the Action Command's Report, contained in NOU 1977:57 A, at p. 38. Unofficial English translation.
2. Ibid.
3. See St. meld No. 25 (1973-4) at p. 32. Unofficial English translation.
4. See Safety Regulations of 9 July 1976, s. 44.
5. On civil liability see NOU 1973:8; Fleischer in (1976), 20, *Scandinavian Studies in Law*, pp. 105 *et seq.*; and *infra* at note 27 *et seq.*
6. Information given in NOU 1977:57 A, at p. 7, etc.
7. Ibid.
8. NOU 1977:57 A, p. 7.
9. St.prp. (proposition to the Storting) No. 182 (1975-6).
10. NOU 1977:57 A, at p. 8.
11. NOU 1977:57 A, p. 11. Observations on the oil's extent during the period 23 April to 4 May, with maps, are found at pp. 40-3.
12. See IKU Report No. 90 (see above) of September 1977, quoted in NOU 1977:57 A, at pp. 38-9.
13. NOU 1977:57 A, p. 39.
14. NOU 1977:57 A, p. 9.
15. NOU 1977:57 A, at p. 48.
16. *Northern Offshore*, Vol. 6, No. 5 (May 1977), p. 6.
17. Ibid., pp. 6 and 7.
18. Unofficial English translation, provided by the Ministry of Industry (now the Ministry of Oil and Energy).
19. Ibid.
20. See NOU 1977:57 A, pp. 58-60. Phillips would have preferred to use chemicals to a larger extent than allowed by the action command.
21. Report of 10 October 1977 by the Commission of Inquiry, appointed by Royal Decree of 26 April 1977, p. 110. Unofficial English translation. For the

Norwegian original, see NOU 1977:47, p. 60.

22. Innst. S.No. 61 (1978-9); cf. also the Government's Report to the Storting, St.meld. No. 65 (1977-8).

23. Report of the Commission of Inquiry (cf. note 21), at pp. 106-7 (NOU 1977:47, p. 58). DSHV = Down-hole safety valve; BMH plug = Baker Model BMH Bypass Blanking Plug with removable Mandral; BOP = blow-out preventer.

24. See note 21.

25. Innst. S. No. 61 (1978-9), at p. 10.

26. Ibid., at p. 14.

27. See note 5.

28. See thus a minority in the Commission on liability set up by the Norwegian Government to consider the question of legislation on off-shore oil pollution liability; see NOU 1973:8, pp. 26-7.

29. Cf. NOU 1973:8, pp. 25 *et seq.* (majority view, unlimited liability) and Fleischer, 'Liability for Oil Pollution Damage resulting from Offshore Operations', (1976) 20 *Scandinavian Studies in Law*, pp. 116 *et seq.*

30. See, *inter alia*, the 1958 Geneva Convention on the Continental Shelf, Art. 2; cf. also the ICNT (Informal Composite Negotiating Text) of the 3rd UN Conference on the Law of the Sea, Art. 77.

31. Geneva Convention Art. 3; ICNT, Art. 78.

32. This view is expressed more fully in my article, 'The Right to a 200-Mile Exclusive Economic Zone or a Special Fishery Zone', 14 *San Diego Law Review* 1977, pp. 548 *et seq.* For a different opinion see Phillips, 'The exclusive economic zone as a concept in international law', 26 *I.C.L.Q.* 1977, pp. 585 *et seq.*; see in particular note 1, p. 585, where it is stated that 'the concept of an exclusive zone . . . has no present standing in international law.'

33. For a more comprehensive review of the different bases of international law which may here be relevant according to the view of the present writer see Fleischer, 'International Aspects of Marine Pollution (pollution from ships and drilling rigs, oil, dumping of waste, nuclear waste)' in *New Directions in the Law of the Sea*, Vol. III, pp. 78 *et seq.*

34. 3 R.I.A.A. p. 1905 *et seq.*

35. Cf. ICNT, Art. 209.

36. See the *Bravo Report*, NOU 1977:47 (and the unofficial translation in 47 A) and *supra*, at notes 21 to 23.

37. See *Conference on Safety and Pollution Safeguards in the Development of North-West European Offshore Mineral Resources* (Summary Records, Department of Trade and Industry, Petroleum Division, 1973). At this Conference Working Groups were set up both for the aspects of safety and for that of civil liability. While the latter part of the work has been finished, as parts of the preparation which led to the 1976 London Convention, the work on safety standards is not yet completed.

6 THE ROLE OF INSURANCE AND REINSURANCE

Samir Mankabady

1 INTRODUCTION

The off-shore and on-shore operations for the exploration and exploitation of oil and its transport require very expensive insurance cover. A
few years ago, the insurance market could only offer a cover of around
$300 million on a single off-shore structure. Nowadays, the value of
certain platforms in the North Sea may well exceed $750 million. Any
platform needs ancillary services such as tugs, modern supply ships,
launch barges, pipe-laying barges, fire-fighting ships, etc. Thus, the large
size of the insurance cover and the limited capacity of world insurance
market is the first feature in the production of oil.

The second feature is the hazard associated with the production and
transport of oil, mainly marine pollution. When thousands of tons of
oil enter the sea,[1] the drifting oil[2] may cause considerable damage to
fishing grounds and the beaches. Cleaning up the beaches and regaining
control of the well are costly operations. Three years ago, it was difficult
to insure pollution damage from off-shore installations simply because
such cover was not available in the market. Today, pollution damage
can be insured, but the premiums and deductions are high.

There are various factors which cause oil pollution. In the case of
tankers, it may be caused by accidents, groundings, collisions, cleaning
and ballast operations and spills in connection with loading, discharging
and bunkering. What is giving much concern to shipping and insurance
circles is pollution caused by tanker operations. It is well known that
for stability and manoeuvring purposes, most tankers have to take ballast
water in uncleaned cargo tanks after loading. This mixture could be
treated through certain systems such as the Load-on-Top (LOT) and the
segregated ballast tank (SBT).

In the case of an oil rig, oil may enter the marine environment as a
result of: (a) blow-out or seepage from the well; or (b) accidents involving the rig, pipelines or off-shore oil storage tanks.

Whatever the cause of pollution, an insurance cover may be required
by statute, or in the terms of a contract, or to avoid extensive claims
which may throw the assured out of business.

Various policies could be effected to cover construction risks, production
and storage against fire, loss of profits, third party liability, workmen's

compensation and employer's liability. However, this chapter is limited to the discussion of insurance policies required in the operation of an oil tanker and an oil rig.

2 INSURANCE FOR OIL TANKERS

In view of the increased potential hazards from oil spills caused by accidents involving tankers, certain safety measures are agreed.[3] Some of these measures aim at preventing accidents at sea and, consequently, at eliminating oil pollution, while others are adopted to raise the standard of safety,[4] as a large number of accidents occur due to human error. Other measures are approved to establish liability and to ensure that adequate compensation is available to persons who suffer damage. Thus, before considering the types of insurance effected in the carriage of oil, it is necessary to outline the conventions, establishing liability and those limiting such liability.

2.1 The Conventions on Liability for Oil Pollution Damage and on the Limitation of Liability

The important conventions in this respect are:

 (i) the Convention on Civil Liability for Oil Pollution Damage (CLC), 1969;
(ii) the Convention on Limitation for Maritime Claims, 1976; and
(iii) the Fund Convention, 1971.

2.1.1 The Convention on Civil Liability for Oil Pollution Damage (CLC) 1969

Under this Convention, the owner of a tanker, subject to a number of exceptions, is absolutely liable for any damage caused by oil pollution. It, in fact, imposes strict liability on tanker owners for marine pollution. It also requires (Art. VII) that owners of tankers carrying more than 2,000 tons of oil in bulk as cargo should maintain insurance or other financial security. The State of registry is required to certify that this requirement is complied with. An oil tanker would not be able to trade with Convention countries without carrying such certificate. Further, the Convention gives a right of direct action against the insurer or the guarantor.

Limitation of Liability under the CLC. Art. V(1) states: 'The Owner of a ship shall be entitled to limit his liability under this Convention in respect of any one incident.' Therefore, it is only the liability under the Convention which the owner can limit in accordance with Art. V. That liability is, by Art. III(1), 'for any pollution damage caused by oil which has escaped or has been discharged from the ship as a result of the incident'. Consequently, the following types of liability are not covered by the limitation system provided for by the CLC:

(i)   oil escaping from off-shore installations;
(ii) oil escaping from dry cargo ships and tankers not carrying oil in bulk as cargo;
(iii) damage suffered by installations outside the territory or territorial sea of a contracting State and damage suffered on the territory or territorial sea of a non-contracting State; and
(iv) claims against salvors and bareboat charterers.

By Art. V(1), the owner may limit his liability in respect of any one incident to 'an aggregate amount of 2,000 francs for each ton of the ship's tonnage. However, this aggregate amount shall not in any event exceed 210 million francs' (about $14 million).[5] The owner will lose the right of limitation if the incident occurred as a result of his actual fault or privity (Art. V(2)).

It must be noted that, while the owner can limit his liability under the CLC according to the system stipulated in it, other parties, such as a demise charterer, can invoke the 1957 Brussels Convention and limit his liability 'in respect of claims arising from' occurrences of property damage.[6]

Claims for Oil Pollution Damage in the United States. The CLC has not been ratified by the United States and the only two oil Conventions in force are the 1954 Convention on the Pollution of the Sea by Oil and the 1969 International Convention Relating to the Intervention on the High Seas in Cases of Oil Pollution Casualties. Therefore, recovery for oil pollution damage may seem to be easier in the United States, which might be considered as 'forum shopping'. The action brought by the French Government in September 1978 before the US District Court of New York City against Amoco International in respect of the *Amoco Cadiz* stranding may support this view.

In 1970, the Water Quality Improvement Act (WQIA) was passed to cover the federal clean-up costs. Under it, owners or operators of vessels discharging oil are liable to the Federal Government for costs of clean-up to the extent of the lesser of $14 million or $100 per gross ton. This Act was modified by the Federal Water Pollution Control Act, 1972.

When the State of Florida passed the Oil Spill Prevention and Pollution

Control Act, 1970 (a State Act), it was challenged by the shipping interests in *Askew* v. *American Waterways Operator*.[7] The District Court held that the State Act was unconstitutional as it conflicted with federal maritime law-making powers. However, the Supreme Court reversed this decision,[8] because the State Act did not provide for unlimited liability for clean-up or other damages. Consequently, it was construed to be consistent with federal law.

This decision paved the way for many States to have their own Acts beside the federal Act. Since the federal Act does not protect private claims nor claims by the State for clean-up costs, the scope and the protection afforded by this Act and the State Acts differ. However, the interaction of these Acts with the Limitation Act of 1851 give rise to difficult problems.

The following example illustrates how these Acts work together. Suppose a ship of 10,000 tons caused pollution damage and the cost of the federal clean-up reached $1.5 million. Under the WQIA, the vessel is liable to the extent of $1 million ($100 per ton). If the value of the vessel after the incident is $400,000 and the anticipated freight is $100,000, this half million dollars will make up the fund for the other injured parties. Thus, the federal clean-up claims are the only claims protected against the limitation (within its ceiling), as the WQIA and limitation are fully independent.[9]

2.1.2 The 1976 Limitation Convention

This Convention is not yet in force but, when it becomes effective, it is designed to replace the International Convention Relating to the Limitation of the Liability of Owners of Sea-Going Ships, 1957.

Art. 1 of the Convention enumerates the persons entitled to limit their liability. Claims for oil pollution damage are covered by the provisions of the Convention. However, there are three main exceptions relating to the CLC, drill ships and off-shore installations.

(i) Art. 3(e) provides: 'The rules of this Convention shall not apply to . . . (b) claims for oil pollution damage within the meaning of the International Convention on Civil Liability for Oil Pollution Damage, dated 29 November, 1969 or of any amendment or Protocol thereto which is in force.'

Thus, all claims for oil pollution damage as defined by the CLC are excluded from the limitation system of the 1976 Convention. As mentioned earlier, oil pollution damage outside the territory or territorial sea of a contracting State and all damage suffered on the territory or territorial sea of a non-contracting State are not governed by the CLC.

This is so, even if the action for compensation for such damage is brought before the courts of a State party to both the 1976 Convention and the CLC. In such a case, the liability will not be subject to limitation under either Convention. Furthermore, in the case of a claim for compensation brought in a non-contracting State to the 1976 Convention, as for instance the claim of the French Government in the *Amoco Cadiz* in New York, it will not be subject to the limitation system provided for by the 1976 Convention. In brief, in cases where oil pollution damage, as defined in the CLC, would not result in a liability governed by this Convention, the claims will not be subject to limitation either under the 1976 or the CLC.

(ii) Art. 17(4) states:

The Courts of a State Party shall not apply this Convention to ships constructed for, or adapted to, and engaged in drilling: (a) when that State has established under its national legislation a higher limit of liability than that otherwise provided for in Art. 6; or (b) when that State has become party to an international convention regulating the system of liability in respect of such ships.

(iii) Art. 17(5) (b) excludes floating platforms constructed for the purpose of exploring or exploiting the natural resources of the sea bed or the sub-soil thereof.

Under Art. 4:

A person liable shall not be entitled to limit his liability if it is proved that the loss resulted from his personal act or omission, committed with the intent to cause such loss, or recklessly and with knowledge that such loss would probably result.

The amount of limitation is calculated on a sliding scale. For a ship with a tonnage not exceeding 500 tons, the amount is 167,000 SDRs (about £117,647). For a ship with a tonnage in excess of 500 tons, the following amounts must be added to the above figure:

for each ton from 501 to 30,000 tons 167 SDRs (about £118);
for each ton from 30,001 to 70,000 tons 125 SDRs (£88); and
for each ton in excess of 70,000 tons 83 SDRs (£59).

2.1.3 The Fund Convention, 1971

At the end of the International Legal Conference on Marine Pollution
Damage held in Brussels during 10-29 November 1969, a Resolution
was adopted requesting IMCO to convene an international legal conference for the consideration and adoption of a scheme to compensate
victims of oil pollution unable to obtain full compensation under the
CLC. Such a conference was held in Brussels from 29 November to 18
December 1971, and resulted in the adoption of an International Convention on the Establishment of an International Fund for Compensation
for Oil Pollution Damage.

Under the CLC, there are two situations where the victim of oil
pollution may not receive full compensation: first, where the incident
arises out of one of the exceptions mentioned in Articles 3(2) and 11
(war, failure of a government to maintain lights, etc.) and so the owner
is not liable at all; and second, where the amount of damage suffered
exceeds the limits of the tanker-owner's liability. The Fund Convention
remedies this situation by establishing an International Oil Pollution
Compensation Fund. The victim of oil pollution coming in one of the
two categories mentioned above, or who is unable to obtain full compensation because the shipowner is unable to meet his financial obligations
arising out of the CLC, is to be compensated out of the Fund. Even here
his compensation is limited, the maximum being US $30 million.[10]

The second major purpose of the Fund is to relieve shipowners of
the additional financial burdens placed on them by the CLC which
limits the shipowner's liability at US $134 per ton to a maximum of
US $14 million. The Fund is under no obligation to pay where the
pollution damage results from the wilful misconduct of the owner, or
where it results from the failure to observe the provisions of certain
Conventions concerned with the safety of shipping or oil pollution,
where such failure is the cause of the damage.

2.2 Types of Insurance Policies Required

Oil is usually carried under charter parties which may fall under two
main categories:

(i) voyage charter party concluded through a 'spot' fixture; or
(ii) time charter party which may be for a certain period or for a
     series of consecutive voyages.[11]

In view of the special nature of the cargo, the tankers, the hazards
and the conditions of the oil market,[12] charter parties for oil have their

own special terms. Furthermore, the introduction of the two schemes TOVALOP and CRISTAL[13] led to the insertion of certain clauses for the apportionment of liability between the tanker-owners and the cargo-owners for oil pollution damage.

Thus, certain risks are covered by policies taken by tanker-owners, while others are taken by cargo-owners. Beside the insurance policies, the assured may ask the underwriter to provide the following services:

(i)   an insurance analysis (risk management) and loss control and production services; and

(ii)   an insurance programme which is compatible with the international and national regulations in force.

In order to fix the premium, the underwriter had to classify and grade the risk and to estimate the size of a probable maximum loss. He also takes into account the assured's experience in handling the oil and his previous records. For this purpose, the engineering reports are carefully studied.

Usually, risks of this magnitude are reinsured on a treaty basis, and the reinsurers are involved from the beginning in the negotiation for the insurance cover.

2.2.1 The Owner's Insurance

Hull and Machinery Insurance. The insurance on the physical loss, or damage to the tanker, is customarily known as 'Hull and Machinery Insurance'. A hull policy covers any damage or loss caused by certain events such as perils of the sea or negligence of the crew. To such a policy, the Institute Time Clauses (Hull) will be attached. If the policy is effected in the United States, the American Institute Time (Hulls) clauses would be used. Of particular importance is the clause on the trading area, which defines precisely the area where the vessel would be employed. This clause usually follows the Institute Trading Warranties.

The American Hull Syndicate has recently come forward with a proposal for additional insurance cover under hull policies for the risk of a vessel being destroyed by governmental authority acting under statutory powers granted by anti-pollution legislation following a casualty by marine or war perils. These powers are usually available where a substantial threat to public health or welfare has arisen and the destruction of the vessel under such powers does not constitute a claim on the hull policy in its usual form.

For an additional premium, the syndicate is prepared to extend policies in these terms:

> This insurance also covers loss of or damage to the vessel directly
> caused by governmental authorities acting for the public welfare to
> prevent or mitigate a pollution hazard, or threat thereof, resulting
> directly from damage to the vessel for which the underwriters are
> liable under this policy, provided such act of governmental author-
> ities has not resulted from want of due diligence by the assured, the
> owners, or managers of the vessel or any of them to prevent or
> mitigate such hazard or threat. Masters, officers, crew or pilots are
> not to be considered owners within the meaning of this clause should
> they hold shares in the vessel.

War and Related Risks Insurance. The second area for physical damage
insurance is against war risks. Traditionally such risks have been insured
separately because different considerations may be taken into account
in deciding the unstable areas.

The war risk policy covers perils taken out of the standard form of
the marine policy by the F.C. & S. clause. Certain specific perils which
are not covered by the word 'war' are added in the policy, for example
hostilities, warlike operations, etc.

Loss of Earning Insurance. A cover exists in the insurance market for
oil tankers employed under a time charter party which is known as 'loss
of earning' or 'loss of hire' insurance. Under it, the insurer will provide
'off hire' coverage up to a specified number of days. It could be 90 days
or even 180 days, after a waiting period of a specified number of days
in each incident (usually between 14 and 30 days).

Protection and Indemnity (P. & I.) Cover. The protection and indemnity
cover is placed with a Mutual Association of Owners, with a part of the
total risk of the Association placed in the insurance market. The P. & I.
Clubs work on the basis of a preliminary insurance rate and a supplementary 'call' at the closing of the accounts. The P. & I. Clubs offer coverage
against various risks such as: loss of life; personal injury; medical expenses
of the crew; repatriation of the crew; strikes; removal of the wrecks
where legally required; damage to fixed objects (for example to jetties
or navigation beacons); fines and penalties imposed by port authorities;
and legal costs of litigation.[14]

More important is the cover offered for oil spillage under the Tanker
Owners' Voluntary Agreement Concerning Liability for Oil Pollution

(TOVALOP). Tanker-owners set up this scheme which was sponsored by seven oil companies.[15] The purpose of the scheme is to deal only with compensation claims by national governments or any public authority, for example a local authority or a coast guard. By signing TOVALOP a tanker-owner undertakes to pay any government compensation for pollution damage up to $16.8 million per tanker. For this reason, he must have an insurance cover for this sum.[16] Such an insurance can be provided by a P. & I. Club, the International Tanker Indemnity Association Ltd, or any insurer acceptable to the International Tanker Owners' Pollution Federation Ltd.

TOVALOP came into operation in October 1969, and it was designed to fill a gap in national and international legislation so that governments and tanker-owners could recover the costs of cleaning up, and to provide a uniform solution on an international level to the problem of oil pollution clean-up compensation.

It provides that in case of pollution:

(i)   the owner of the tanker will either remove it or reimburse the government concerned for reasonably incurred clean-up costs. In case of dispute on the measures or the costs, such dispute will be resolved by arbitration;

(ii) the owner assumes liability for clean-up costs unless he could show he had not been at fault;

(iii) the liability is limited to US $160 per ton, subject to a maximum of $16.8 million for each TOVALOP tanker involved in any one incident.[17]

2.2.2 The Cargo-Owner's Insurance

Owners of oil tankers felt that it was not fair that they carry all the burden in case of oil pollution and that the cargo-owners, i.e. the oil industry, should share in this respect. Further, TOVALOP does not provide a sufficient cover. For these reasons, a scheme called the Contract Regarding an Interim Supplement to Tanker Liability for Oil Pollution (CRISTAL) was adopted. According to this scheme, owners of oil cargoes provide the means for compensation to any one who suffers oil pollution damage. They will pay compensation to individuals, for example fishermen, hoteliers, etc., as well as to governments. The scheme also increased the amount payable for compensation for a pollution incident to a limit of $30 million by paying the difference between the TOVALOP liabilities and the actual loss or expense.

Claims agreed by CRISTAL are paid from money subscribed by its members. The share of each member is fixed according to his share of shipments of crude and fuel oil.[18] Thus, TOVALOP is the tanker-owners' scheme and is concerned with clean-up costs. The insurers are mainly the P. & I. Clubs. CRISTAL is an oil industry scheme for clean-up and third party damages, and the insurers are the cargo-owners themselves.

3 INSURANCE FOR OFF-SHORE OPERATIONS

The risks in off-shore operations which need to be covered by insurance could be described as political, economical, technical and operational. Obviously, the insurance cover depends on many factors such as the type of the rig, the operating area, the stage reached in the operation and the terms of the drilling contract and other contracts where additional liability is required.

Where there is a possibility of expropriation, confiscation or a change in the terms of the operating agreement, the licensee would cover such political risk by an insurance policy. In case there are changes in the price of oil or gas during the life of the field, such economical risks could also be insured.

The technical risks are mainly related to the geological hazards, and the operational risks are concerned with pollution, blow-out,[19] fire and other accidents. Other risks could derive from the design of the installations, the failure of the equipment, and towing the rig from the shipyard to the site.

Usually the licence (the operating agreement) is given to a subsidiary or a company which is an offshoot of a multinational oil company. The licensee is called the operator and he may own the rig (normally when it is a fixed platform) or hire it (when it is a moveable unit). The operator entrusts drilling to a contractor who may be at the same time the rig-owner. The drilling contractor engages for all or certain parts of the contract subcontractors. Thus, we have the operator and the contractor who may or may not be the rig-owner and the liability of each party is determined, in principle, according to the terms of his contract and the law of the State issuing the licence.

As with oil tankers, the first section of this chapter deals with the convention establishing liability for oil pollution damage and the second considers the types of insurance policies required.

3.1 The London Convention on Civil Liability for Oil Pollution Damage from Off-shore Operations, 1976

In December 1976, the above-mentioned Convention was signed in London.[20] Its main purpose is to set up a legal regime for off-shore operations similar – to a certain extent – to the one established by the CLC.[21]

Article 3 of the Convention provides that the operator is liable for any pollution damage resulting from the installations which are defined as 'any well or other facility, whether fixed or mobile, which is used for the purpose of exploring for, producing, treating, storing, transmitting or regaining control of the flow of crude oil from the seabed or its sub-soil'. Abandoned wells are covered for a period of five years after they have been plugged off.[22] Any ship or barge used exclusively for storage purposes on the site would also come within the definition of an installation.[23] As for the term 'pollution', it covers damage caused by contamination and cost of preventive measures, but it does not cover damage caused by fire or explosion.

Like the tanker-owner, the operator can exclude his liability if he proves that the damage resulted from an act of war, insurrection and natural phenomenon. He cannot, however, exclude his liability for acts of sabotage.[24]

3.1.1 Limitation of Liability

The insurance cover available in the market was the central point for discussion when the draftsmen of the Convention considered the limits[25] of the operator's liability. Some delegates were in favour of fixing the limit according to the cover obtainable in the market. However, there was no agreement on the amount of cover available; it could be $20 million and it may even reach $80 million. Other delegates believed that the operator's limit should be higher than the insurance cover. The difference between what the underwriter will pay and the operator's limit will be justified by the industrial risk taken by the operator. This approach prevailed and Art. 6 fixes the limit at 30 million Special Drawing Right (SDR), and the insurance cover for 22 million SDR for the first five years commencing from 1 May 1977; that is the date on which the Convention was opened for signature. After this period the limit will be 40 million SDR, and the insurance cover required from the operator is for 35 million SDR.[26]

It must be noted that a higher limit could be imposed. Art. 15 states:

> This Convention shall not prevent a State from providing for un-limited liability or a higher limit of liability than that currently applicable under Art. 6 for pollution damage caused by installations for which it is the Controlling State and suffered in that State or in another State Party; provided however that in so doing it shall not

discriminate on the basis of nationality. Such provision may be based on the principle of reciprocity.

Also the Controlling State has the right to specify the type and the terms of the insurance cover or any other financial security which would be acceptable to her.[27]

3.2 Types of Insurance Policies Required

The insurance requirements for off-shore operations differ from one stage to another. The pre-drilling stage presents no difficulty except when the licence is granted to more than one company. The one leading the operation is called the operator and the agreement between the companies will determine the liability and the insurance cover required.

In the next stage of exploration,[28] the licensee may hire the rig, charter supply vessels[29] and enter into agreement with the drilling contractor. All these agreements contain provisions on indemnity and the insurance cover required. Sometimes the licensee takes an insurance policy as a cover for any risk left by him or his contractors.

The third stage is the construction and the installation of the rig. The main problem here is that during the shipyard building the rig may rely on equipment to be supplied by different contractors. Any delay in delivery by these contractors would lead to a delay in delivery of the rig and, consequently, the commencement of production. Whatever the provision on delay in the shipbuilding contract, the consequences of delay could be covered by an insurance policy, credit insurance, or by a governmental guarantee.

For moving the rig from the yard to the site, a contract of towage or perhaps a contract of carriage[30] will be needed. As this is one of the hazardous operations,[31] an insurance cover is taken. The policy usually states: 'warranted approval of tug, tow, towage and stowage arrangements . . . (the name of the surveyor or his firm) and all recommendations complied with'. Although the surveyor[32] is approved and often appointed by the underwriter, he is employed by the rig-owner. Obviously, one of the most important factors in the success of the towage or the carriage operation is the weather. Consequently, the insurer imposes another strict warranty on sailing in the policy.

Once the rig successfully reaches its site, 24 or more wells are drilled from a single platform.[33] Drilling may cause pollution, blow-out, personal injuries or damage to the rig itself. Thus, the main areas for cover are:

(i)   pollution and seepage insurance;

  (ii)   blow-out (and cratering);
 (iii)   employer's liability;
 (iv)   the rig; and
  (v)   P. & I. third party liabilities.[34]

3.2.1 Pollution and Seepage Insurance

Usually the licence provides that the licensee is liable for all kind of
damages in connection with his activities. In case the licensee is a subsidiary company, the parent company would be asked to act as a guarantor.
Where there is more than one licensee in one block, they are all considered
jointly liable for the damage. The Petroleum (Production) Regulations
1976 provide for Model Clauses to be included in the licences, providing
that the licensee is responsible for the prevention of pollution. In the
case of a massive pollution, the licence may be revoked.

   Thus, the licence and the Hold Harmless Agreement make the licensee,
i.e. the operator, liable for pollution damage.[35] Whether such liability is
strict or based on negligence, excluding damages caused by natural
phenomena, depends on the terms of the licence and the law of the State
issuing it.

   To cover such liability, the operator effects a policy called 'Pollution
and Seepage' to pay claims for cleaning-up and other costs. The cover[36]
extends the protection to contractors and the subcontractors. However,
the pollution and seepage policy contains a number of exclusions and
fines, penalties, the costs of regaining control on the well and damage to
the assured's property.

   Sixteen oil companies made an agreement called the 'Offshore
Pollution Liability Agreement' (OPOL), which came into effect on 1 May
1975.[37] The Agreement provides for setting up an association whose
members will accept strict liability to pay for pollution damage and/or
remedial measures arising from an oil spillage incident attributable to
their off-shore operations.

   The agreement is designed to cover the off-shore waters of any State.
It applied initially to the UK continental shelf and was later made
applicable to other North Sea States; it can be extended to any other
State which the members by appropriate amendment to the Agreement
may nominate.

   Claims for pollution damage and remedial measures will be met by
the operator on behalf of those companies for which he is operating. In
this respect, pollution damage means direct loss or damage by contamination which results from a discharge of oil. Remedial measures are
reasonable measures taken by any member from whose off-shore facility

the discharge of oil occurs, and by any State to prevent, mitigate or eliminate pollution damage. The maximum liability is $25 million per incident. This is made up of $12.5 million to cover pollution damage claims and $12.5 million for remedial measures. When all claims in one category have been met, any surplus may be used to meet the claims in the other category. Parties to OPOL undertake to establish and maintain financial security for an amount of $25 million for any one occurrence and $50 million in the annual aggregate.

There are exceptions to the strict liability principle, as no obligation shall arise with respect to pollution damage and remedial measures taken if the incident:

   (i)   resulted from an act of war or a natural phenomenon of an exceptional, inevitable and irresistible character;

  (ii)   was wholly caused by an act or omission done with intent to cause damage or was wholly caused by the negligence or other wrongful act of any State or other authority;

 (iii)   resulted wholly or partially, either from an act or omission done with intent to cause damage by a claimant, or from the negligence of that claimant.

The agreement will continue in effect for a period of six years and thereafter from year to year, and will be governed by English law. When the London Convention 1976 comes into force, OPOL would have to be brought into line with the Convention.

3.2.2 Blow-Out (and Cratering)

Blow-out is the most important risk and may happen as a result of various reasons, such as hitting a high-pressure oil reservoir, or an accident caused by heavy weather, fire, explosion or human error in handling the blow-out preventer (BOP).[38]

In the case of a blow-out, the drilling contractor would be the person liable for any pollution damage. He would also be liable if the blow-out was caused by negligent acts or omissions on the part of his employees. However, such a situation would prevent drilling contractors from carrying out their functions. Therefore, in drilling contracts, the 'Hold Harmless Agreement' is incorporated. According to this agreement, the operator holds the drilling contractors harmless of all claims resulting from pollution damage.[39]

As a result, the operator needs another policy; a Blow-Out (and Cratering)[40] to cover the costs incurred in regaining control of the well in the event of uncontrolled blow-out and/or cratering which is excluded in the Pollution and Seepage policy. It is warranted that the assured uses every endeavour to ensure the fitness of the blow-out preventers and other equipment, and to control the well, or to stop the escape of oil in the event of a blow-out.[41]

3.2.3 Employer's Liability

An employee has the right to work in an environment reasonably free from dangers.[42] Therefore, governments in most countries impose compulsory insurance to cover the employer's liability. In the UK, the application of the Employer Liability (Compulsory Insurance) Act 1969 has been extended to off-shore operations by two sets of statutory regulations: the Offshore Installation (Application of the Employer Liability) (Compulsory Insurance) Act 1969 Regulations 1975 and the Employer's Liability (Compulsory Insurance) (Offshore Installation) Regulations 1975.[43] Under these Regulations, a cover for a minimum of £2 million is required to meet claims for bodily injury or disease sustained by an employee.[44]

3.2.4 Insurance Covering the Rig

The rig-owner takes an insurance covering the rig itself. There are three different types of rigs: the jack-up, the semi-submersible and the drill ship. In addition, the submersible barges and fixed platforms can be used in shallow waters.[45]

It must be noted that the Mineral Workings (Offshore Installations) Act 1971 applies to all off-shore installations without any distinction (s. 1). However, if the rig is considered to be a ship, it will also be subject to the provisions of the 1894 Act, and consequently the rig-owner will be entitled to the limitation of liability according to the 1958 Act.

Insurance policies on rigs were afforded coverage by either the Institute Time Clauses – Hulls or American Institute Hull Clauses. Since March 1972 the London Standard Drilling Barge Form – All Risks started to be widely used. In practice, the Hull Clauses are used for rigs to be considered as ships, for example self-propelled rigs and the drill ship, while the Barge Form is used for jack-up and semi-submersible rigs.[46] Thus, there are two types of policies: the Hull Clauses[47] and the Barge Form. In addition, a war risk policy covers the risks which are excluded from the marine policy[48] by the F.C. & S. clause.

The Hull Clauses. The nature of the exploration and exploitation operations led to the appearance of two important clauses in the Hull form.

(i)   A clause restricting the movement of the drill ship or the self-propelled rig and limiting it to certain areas; for example the Celtic Sea. Usually, the Hull policy on a ship does not contain any restrictions on her movements. Occasionally there may be a warranty excluding certain areas, for example places where ice in winter could be dangerous for the safety of the ship. For rigs considered as ships, drilling is the primary purpose and therefore a clause would be needed to limit the area where drilling could take place.

(ii)   A clause that the drill ship or the self-propelled rig is classed by an approved classification society and that this class would be maintained during the term of the policy.

The London Standard Drilling Barge Form. It may be useful to summarise the important clauses in the London Standard Drilling Barge Form. *Period of insurance*: The period of insurance is usually one year and therefore this Form is a time policy. The Hull Clauses are, in fact, a combined time and voyage policy as Clause 4 provides:

Should the vessel at the expiration of this policy be at sea or in distress or at a port of refuge or of call, she shall, provided previous notice be given to the underwriters, be held covered at a *pro rata* monthly premium to her port of destination.

The particulars of the property insured: The Form provides that the policy covers the hull and machinery, equipment, tools, caissons, materials, derricks, drill stem, casing and tubing while aboard or on vessels moored alongside and all scheduled property owned by or in the care, custody or control of the assured, except those excluded in the policy.
Navigation limits: The limits for the movement of the rig would be stated expressly in the Form. Towage within the specified limits is covered. Also, the Form covers up to 25 per cent of the sum insured any part of the property when separated from the rig whilst temporarily stored or in local transit, provided it is within the specified limits.
Coverage: The Form covers, subject to the exclusions, all risks of direct physical loss or of damage to the property insured so long as such loss or damage has not resulted from want of due diligence by the Assured, the Owners or Managers of the rig.
Collision liability: The clause on collision liability in the Form is rather similar to one used in the Hull Clauses. Exclusions under this clause are removal of wreck, injury to real or personal property, pollution, loss of

life and personal injury or illness.

It is rather interesting to note that the collision clause in the Barge Form uses in the description of the subject-matter insured the word 'vessel'. In fact, the Offshore Installations (Registration) Regulations 1972 mentioned two types of installations, namely the mobile and fixed installations (Reg. 2.1). However, this distinction is adopted only with regard to the particulars required for the registration. Therefore, it could be argued that the jack-up or the semi-submersible are not vessels. Consequently, the operator of such a rig would not be able to rely on the 1958 Act.

Deductible: The Form provides for an amount, left in blank to be agreed between the parties, to be deducted from all claims, but this would not apply in case of total loss.

Exclusions: The Form contains a long list of exclusions, such as earthquake, volcanic eruption, fire, explosion, tidal wave consequent upon earthquake or volcanic eruption. Also excluded is loss or damage caused by or resulting from delay. Damage resulting from drilling of relief wells associated with other units is only covered if notice is given to the insurer and if required an additional premium should be paid. Property used in controlling or attempting to control blow-out, or cratering associated with blow-out, is not covered. Removal of debris is excluded. Loss or damage to the drill stem underground or underwater is only covered if it is a direct result of fire, blow-out or cratering or total loss of the drilling barge caused by a peril insured against. Also excluded from the cover afforded by the Form is well(s) and/or hole(s) while being drilled.

Constructive total loss: In order that the loss be considered as constructive total loss, the expenses of recovering and repairing the rig should exceed the insured value.

Sue and labour expenses: The expenses under this clause would cover up to 25 per cent of the insured value of the items in respect of which such expense is incurred.

Lay-up and cancellation: The premium would be returned to the assured for any period of 30 or more consecutive days of which the rig is laid up in port unemployed. There is also a provision on the return of premium in case of cancellation of the policy.

Time limit: The time limit for bringing an action against the underwriter is two years.

Confiscation and War Risks. Both the Hull Clause and the Barge Form exclude war risks by the F.C. & S. clause. Therefore, the operator would effect a war risk policy which is similar to the ordinary war risk policy. A confiscation policy is a guarantee that if the host government takes away the right of ownership, by confiscating the property without compensation, or by forcing its sale to local businessmen at a loss, under-writers will pay any shortfall in adequate compensation.

3.2.5 P. & I. Cover

In the past, the policy of the London Group of Clubs was that liabilities arising out of drilling operations would not be acceptable in their pool and reinsurance arrangements. However, this policy has now been reversed and the Group accepts in the pool any ship or barge used for drilling. Units which are not considered to be ships, storage units and wreck removal are still excluded.

Most of the P. & I. Clubs adopt the distinction between the liability of the licensee and the contractor. The latter can obtain the normal P. & I. third party liabilities cover subject to certain exclusions; these relate to risks which are borne by other parties or do not arise out of drilling. The exclusions include: loss of hole or well; loss of or damage to in-hole equipment; cost of control including the drilling of relief wells; product liability; and contractual ocean towage liability. The contractor can insure oil pollution damage arising out of the drilling operation (i.e. other than the sample and bunkers tanks or from collision). As for oil pollution arising from beneath the sea, this is the liability of the operator.

4 CONCLUSIONS

It is usually argued that pollution damage from oil tankers is much more serious than those from off-shore operations. This is so because accidents in tankers occur close to the shore, while off-shore installations may be a hundred miles off the coast. Also, off-shore operators offer more financial security than tanker-owners. However, oil pollution damage is an expensive operation, and a risk which maritime countries have come to regard as unacceptable.

The recent measures adopted by IMCO, especially the Convention for the Prevention of Pollution of the Sea from Ships, 1973 (MARPOL), constitute an important step towards the prevention of oil pollution. When the 1978 Protocol relating to that Convention (MARPOL Protocol) comes into force, it will make a valuable contribution to the elimination of pollution.

As a result of the *Amoco Cadiz* incident, France introduced regulations for navigation in the Atlantic and Channel approaches. The Anglo-French Safety of Navigation Group has also agreed on a scheme for the English

Channel. Finally, the European Maritime Pilots' Association presented proposals for a new compulsory pilotage service for the English Channel, the Dover Straits and the southern North Sea, with a mandatory routing of vessels and a radar surveillance system, to the House of Commons Select Committee on Tanker Safety. The main idea in all these schemes is to introduce a tide control on the movement of oil tankers.

Yet, the limitation of liability system provides shipowners with protection from extensive claims. In the *Torrey Canyon* incident in 1967 the cost of cleaning up exceeded $8 million. In limitation proceedings by the owners, the Court of the Southern District of New York estimated the value of the remaining lifeboat from the vessel at $48. Had the case not been settled, most probably the British and French Governments would have received only $48 for their cleaning-up costs.

Although the concept of limitation is a sound one, the final question seems to be who should bear the burden of loss from oil spillage? At present, a good deal of it is borne by innocent victims and coastal States.

Recently, the United States Senate has voted to establish a $200 million fund to compensate oil spills. The revenue for the fund would be raised through a 3 cent-a-hand fee to be levied on crude oil which arrives at US refineries. The EEC Commission also considered several proposals in this respect, such as holding a black list of tankers causing pollution as a kind of criminal record; the extension of the territorial waters to 12 miles; the setting up of a joint EEC Coast Guard service; and finally to increase the compensation available in case of oil pollution damage. Perhaps the last proposal would prove to be the most practicable solution, as the insurance market would be able to arrange for such an increase.

For off-shore operators, it would be preferable to reach an agreement on an international level to make the licensee the person liable for oil pollution damage, up to a certain ceiling. It is true that with the Hold Harmless Agreement the contractor could make the licensee liable for such damage, but his power in this respect depends on the condition of the market. Oil companies operating in certain areas can set up a pool for oil compensation damage for a certain area. Above this ceiling, it would be appropriate that the liability would rest with the State granting the licence.

NOTES

1. On 16 March 1978, the *Amoco Cadiz* (233,000 tons) ran into sharp reefs

and broke up off the Breton fishing village of Portsall – more than 220,000 tons of oil escaped and caused extensive damage.

2. Drifting oil of unknown origin has also been observed.

3. On the other hand, the oil industry developed the Load-on-Top system whereby tanker washings are collected in a slop tank, and after settling, the water beneath the oil is discharged, leaving only the oil slops. These are subsequently removed or incorporated into the next cargo of crude oil.

The practice of the Load-on-Top and other scientific advances led to the introduction of certain Amendments adopted in 1969. (The required number of ratifications has been reached and the Amendments came into force on 10 January 1978.) Despite the 1969 Amendments, it was felt that there was a need to prevent pollution of the sea by all noxious substances emanating from ships, not only oil. Therefore, MARPOL was adopted.

4. On 16 February 1978, IMCO approved two Protocols: the first relates to the Safety of Life at Sea Convention (SOLAS), 1974; and the second relates to MARPOL and 18 resolutions. See IMCO documents: TSPP/CONF/9; TSPP/CONF/D/1–2–7–8.

According to the MARPOL Protocol, owners of tankers over 40,000 tons would have to operate either clean ballast tanks (CBT), or segregated ballast tanks (SBT) or crude oil washing (COW) systems by 1981. Tankers over 70,000 tons will cease to have the option of CBT by 1983 and those between 40,000 and 69,999 tons by 1985.

Inert gas systems (IGS) will be required on all crude carriers over 70,000 tons by 1983 and for those between 20,000 and 69,999 tons by 1985. The administration of the flag State may grant exemption from an IGS if it is not reasonable and practicable to fit it. Whenever a COW system is used, IGS is mandatory.

As for new tankers over 20,000 tons, they should have SBT in combination with a COW and IGS systems. Tankers over 30,000 tons are required to have IGS and also protective located SBT.

5. A Protocol was adopted in November 1976 which alters the unit of account from the Poincaré francs to the Special Drawing Rights. This Protocol is not yet in force.

6. Under this Convention:

> the owner, charterer, manager and operator of the ship, and . . . the master, members of the crew and other servants of the owner, charterer, manager or operator acting in the course of their employment, may limit their liability for loss of life, or personal injury to, or damage to any other property or infringement of any rights.

Obviously this covers the case of an oil pollution damage.

The limits of liability in case of property claims are 1,000 francs for each ton; 3,100 francs for personal claims and in both personal and property claims is 3,100 francs of which 2,100 francs will be for the payment of personal claims and 1,000 francs for property claims. The sterling equivalent of these figures is declared from time to time in Statutory Instruments, the current one being the Merchant Shipping (Sterling Equivalents) (Various Enactments) Order 1979, S.I. 1979/790.

7. 335 F. Supp. 1241 (M.D. Fla. 1971).

8. 412 U.S. 933 (1973).

9. It is clear that if the owners of an oil-polluting ship are able to limit their liability under the 1851 Act, the recovery by State and private claimants would be affected. However, 'privity and knowledge' could be established easily under American law by, for example, invoking the concept of unseaworthiness, i.e. some defect in the vessel or equipment. In fact, the French Government, by suing in the United States, is hoping to prove the privity or knowledge of Amoco International,

i.e. the body which was effectively in control of the operation, in order to avoid the limitation. This move brought sharp reaction from insurance circles in London, as it is considered to be contrary to Art. III(4) of the CLC, which is ratified by France.

10. The Fund will be made up of about £19.5 million and this could be raised to £39m. It will operate on a mutual basis in a similar manner to CRISTAL.

11. In a depressed market, tanker-owners usually adopt one of the following ways: (i) to fix their ships on long-term charters; (ii) to use the ship to carry dry cargo if this is possible; or (iii) to lay up the vessel.

12. The main customers of tanker-owners are the world's oil companies, i.e. usually groups consisting of a parent company and subsidiaries, which produce, refine and distribute the oil. About seven of these are giants, but there is also a substantial number of smaller ones. Tanker-owners, on the other hand, are more numerous, frequently consisting of 'one-ship' companies or companies owning only two or three ships. See *The 'Siboen'* and *The 'Sibotre'* [1976], 1 Lloyd's. Rep. 293.

13. There are many organisations concerned with tankers and oil pollution. The International Association of Independent Tanker Owners (Intertanko) was founded in 1970. It is a forum for discussion 'of certain specific questions deserving collective attention, without being dealt with by other organisations'. See its *Annual Report, 1973*, p. 2.

Another body is the Institute for Marine Pollution Compensation Ltd, and the International Tanker Indemnity Association Ltd.

14. The P. & I. Club also provides the necessary certificates. As for its advisory role, it includes: (a) liaison with oil companies, national and international organisations concerned with oil pollution; and (b) liaison with the technical department of Tovalop.

15. These seven companies are: Royal Dutch/Shell; BP, Standard (New Jersey); Mobil; Texaco; Standard (California).

16. The administration of TOVALOP is entrusted to the International Tanker Owners' Pollution Federation Ltd, which has its headquarters in London. Some 3,300 tanker-owners are members of the Federation, which also provides a technical advisory service to governments, P. & I. Clubs and to its members on all aspects of clean-up techniques. It keeps records on the various incidents of pollution.

17. In 1972, the International Salvage Union, concerned that liability might attach to salvors for oil pollution caused during salvage operations, decided that their members would not be able to render salvage services unless they were given an indemnity by the shipowner, countersigned by the shipowner's P. & I. Club. As a result, the P. & I. Oil Pollution Indemnity Clause could be incorporated in salvage contracts, where it was in a Club's interest so as to minimise any claims against it. The effect of the clause was to protect the salvor against oil pollution which he might cause. However, in case the pollution damage is caused by the negligence of the salvor, such damage is not covered by the CLC, nor by TOVALOP, which are the terms of reference for CRISTAL. Therefore, negotiations started between the P. & I. Clubs and salvors on a new insurance scheme which came into effect on 20 February 1975. Under this scheme, unit liability of the salvor for pollution is covered up to $20 million per craft with additional limit of $40 million where more than three crafts are involved in one operation. The scheme also provides that where salvors operate as contractors, the insurance cover will be $20 million with a deductible of $50,000.

18. The Oil Companies Institute for Marine Pollution Compensation Ltd collects the shares and pays the claims. Another company in London called Marine Pollution Compensation Services Ltd is entrusted with the day-to-day work of CRISTAL.

19. For the possible blow-outs, see the Study prepared by the Central Unit on Environmental Pollution, Department of the Environment, *Pollution Paper No. 8* (1976).

20. The Convention has been signed by the Netherlands, Norway, Sweden and the United Kingdom, but has not been ratified by any State. The Convention is intended to be regional and applicable only to the North Sea, the Baltic Sea or that part of the Atlantic Ocean to the North of 36° north latitude which could accede to it.

For another regional agreement, see the Barcelona Convention of 16 February 1976 for the Mediterranean. See Abecassis, *Oil Pollution from Ships*, p. 97. See also Fleischer, 'Liability for Oil Pollution Damage resulting from Offshore Operations', (1976) 20 *Scandinavian Studies in Law*, pp. 107-43.

21. For the background of this Convention, see Dubais, 'The 1976 London Convention on Civil Liability for Oil Pollution Damage from Offshore Operations', (1977) 9 J.M.L.C. 28.

22. *Notice to Mariners No. 40 – Notice 2329/78* states:

Suspended wells are those drilled in the exploration for oil or gas but then left (with wellhead gear on the seabed) for possible later use. A suspended well is charted by a danger circle and the legend Well or Wellhead. These wellheads have normally been marked by light-buoys and these are shown on the largest scale charts.

From September 1978 the Department of Energy has released offshore operators from the obligation to buoy suspended wells. Consequently no reliance should be placed on the existence of the charted buoys many of which may already have been removed.

Buoys will not be removed from wells which are:

(a) possible hazards to surface navigation (in particular those near the charted Deep Water Route in the Southern North Sea).

(b) actually in use for the extraction of oil or gas (technically described as 'subsea completions').

23. In the draft Convention on Offshore Mobile Craft prepared by the Comité Maritime International at the Rio de Janeiro Conference in September 1977, craft is defined as

any marine structure of whatever nature permanently fixed into the sea-bed which:

(a) is capable of moving or being moved while floating in or on water, whether or not attached to the sea-bed during operations, and

(b) is used or intended for use in the exploration, exploitation, processing, transport or storage of the mineral resources of the sea-bed or its subsoil or in ancillary activities [Art. 1].

24. If, under the 1969 Brussels Convention, the tanker-owner can exclude his liability for damage which 'was wholly caused by an act or omission done with intent to cause damage by a third party'.

25. The time limit for bringing a claim is twelve months after the date on which the claimant has suffered the damage. However, all actions would be time barred after four years from the date of the incident which caused the damage.

26. The SDR is based on a 'basket' of sixteen major world currencies, and will in turn be converted into sterling at the current market rate. See Silard, 'Carriage of the SDR by Sea; the Unit of Account of the Hamburg Rules', (1978) 10 J.M.L.C. 13.

27. Claims for damage may be brought before the courts where pollution damage has occurred or before the courts of the Controlling State (Art. 11). However, after the fund has been constituted in a party State, only the courts of this State will have to determine the distribution of the fund.

28. Although the exploration phase of the North Sea has now been completed, it is still going on in the West Coast of Scotland, the Irish Sea and the English Channel.

29. He may also use the service of a helicopter.

30. The jack-up drilling rig *Orien* was carried on the derrick barge *Federal 400-2* as deck cargo. The barge stranded off Guernsey but was successfully salved.

31. The last accident – at the time of writing – is the case of the dumb barge *Intermac 600* on which the oil platform jacket *Namorado* was mounted. The *Namorado* platform sank in about 200 feet of water following the parting of tow lines between the *Intermac 600* and the towing vessel. The insurance claim may reach £11 million.

32. The surveyor makes recommendations not only on the rig but also on the tug, the stresses and the equipment. As all this information needs complicated engineering calculations, it is usually fed into a computer.

33. The UK Offshore Operators' Association (UKOOA) and the North Sea Operators' Committee (Norway) (NOSC—N) agreed on co-ordinated plans for dealing with emergencies on North Sea oil and gas installations. The plans divide the North Sea into five geographical sectors. Each sector encompasses a group of fields, and clubs together the various operators into one unit for combating blow-outs, fires and other emergencies. Every sector club has a co-ordinating operator drawn from one of the individual field operators. While each operator will retain responsibility for his own operations and for providing emergency systems, the sector clubs will enable both UK and Norwegian operators to obtain early assistance in case of accidents. See also the Offshore Installations (Emergency Procedures) Regulations, S.I. 1976/1542, which came into force on 18 October 1976.

34. The oil companies agreed to make these plans available to the public. The Hold Harmless Agreement states that: 'The operator shall indemnify, defend and hold the State harmless against all claims, demands and causes of actions.'

35. In this respect, the clause runs as follows:

In the performance of his obligations hereunder, the operator agrees to accept full responsibility for compliance with all applicable laws and governmental orders, rules and regulations relating to pollution such as the Dumping at Sea Act 1974 . . . The operator at his cost and expense will clean up and remove any pollution . . . Notwithstanding the foregoing, the operator shall not be liable for pollution or any seepage emanating from . . .

36. Such cover could be obtained in the ordinary insurance market. Recently, 17 oil companies have started an insurance company in Bermuda called Oil Insurance Limited (OIL) to cover such risks.

37. An Agreement was amended on 12 September 1975 and 23 March and December 1976, and 5 May 1977. See the Booklet entitled *OPOL*, published in September 1977 by the Offshore Pollution Liability Association Ltd.

38. Blow-out preventers (BOP) are devices such as valves which prevent any fire or explosion.

39. In fact, the drilling contractor before commencing his work usually requires certificates indicating: (a) the insurance company or companies offering the cover; (b) waiver of subrogation endorsement has been attached to relevant policies; (c) the territorial limits of all policies; and (d) the commencement and expiry dates of the policies. Furthermore, the operator undertakes to inform the drilling contractor of any change in any policy.

40. In the London Standard Drilling Barge Form, the term 'blow-out' is defined to mean

a sudden accidental, uncontrolled and continuous expulsion from a well and above the surface of the ground of the drilling fluid in an oil or gas well, followed by continuous and uncontrolled flow from a well and above the surface of the ground of oil, gas or water due to encountering subterranean pressures.

The term 'cratering' is defined as a basin-like depression in the earth's surface surrounding a well caused by the erosion and eruptive action of oil, gas or water flowing without restriction.

41. In this respect, products liability questions may also arise.

42. The Health and Safety at Work etc. Act 1974 ensures such an environment. Under this Act, a policy-making body called the Health and Safety Commission was set up. The policies are carried out by the Health and Safety inspectors. See the Offshore Installations (Operational Safety, Health & Welfare) Regulations 1976 which came into force on 15 November 1976 and the Health and Safety at Work Act 1974 (Application outside Great Britain) Order, S.I. 1977/1232, operative 1 September 1977.

43. S.I. 1975/1443 which came into force on 1 September 1975.

44. For jurisdiction over oil pollution emanating from the UK sector of the Continental Shelf, see s. 3 Continental Shelf Act, 1964 and Continental Shelf (Jurisdiction) Order 1968, S.I. 1968/892 as amended by S.I. 1971/721, S.I. 1974/1490, S.I. 1975/1708 and S.I. 1976/1517.

45. In recent years, there have been new designs for rigs. For instance, certain rigs are built on a system which combines a ship-shape hull and a jack-up. According to this system, the vessel is self-propelled and moves between locations as a ship. Also, with its four legs it elevates itself and drills as a jack-up. Perhaps the most revolutionary design is the one which incorporates the principle of a ship, semi-submersible and a jack-up, all in one unit.

46. The Hold Harmless Agreement would not extend to cover the liability of the rig-owner. To achieve such a result an express clause is needed.

47. In case there is a towage operation, the Hull and Machinery Insurance will include the tower's liability clause.

48. The usual clause states:

The rig shall not be bound to proceed to any place and shall not be used in any service which will bring her within a zone which is known to be dangerous as a result of any actual or threatened act of war, hostilities, warlike operations, revolutions, civil war, civil commotion or any acts of piracy or of hostility or malicious damage.

The rig shall not be exposed in any way to any risks in respect of penalties or otherwise howsoever consequent upon the imposition of sanctions by any Government or Governments.

The rig shall not carry any goods that in any way expose her to any risk of seizure, penalties or any other interference of any kind whatsoever by any belligerent or fighting power or party to any *de jure* or *de facto* Government or Ruler.

Part II:
POLLUTION FROM SOURCES OTHER THAN OIL

7 DUMPING AT SEA

Philip Kunig

In addition to pollution as a direct result of navigation and the
exploration and exploitation of the sea bed, the second important
source of marine pollution is the disposal of waste produced on land.
Direct disposal of land-based waste into the sea has to be considered,
but often the method adopted is to transport waste by ship or by aircraft and dispose of it at sea. Disposal of waste at sea, often called 'ocean
dumping', constitutes 10 per cent of the pollutants and toxic agents that
enter the world's oceans.

Regulations for ocean dumping can be found in the Geneva Convention
on the High Seas of 29 April 1958. The matter is exclusively dealt with
by the following agreements: Convention on the Prevention of Marine
Pollution by Dumping from Ships and Aircraft of 15 February 1972
(Oslo Convention) and the Convention on the Prevention of Marine
Pollution by Dumping of Wastes and Other Matter of 13 November
1972 (London Convention). Ocean dumping, as well as other types of
pollution, is considered in the Convention on the Protection of the
Marine Environment of the Baltic Sea Area of 22 March 1974 (Helsinki
Convention), the Convention on the Prevention of Marine Pollution from
Land-based Sources of 21 February 1974,[1] and the Convention for the
Protection of the Mediterranean Sea against Pollution of 16 February
1976.[2]

Also, within the framework of the European Communities and the
Third United Nations Conference on the Law of the Sea (UNCLOS III),
attempts are being made to come to a further improvement and
harmonisation of the protection of the marine environment.[3]

In what follows, a survey is given of the legal situation which led to
the conclusion of the treaties of the seventies; then the Conventions of
Oslo, London and Helsinki are commented on; finally, the proposals
with regard to ocean dumping of UNCLOS III are briefly considered.

1 THE LAW PRIOR TO 1970

1.1

Art. 24 of the Convention on the High Seas is limited to the disposal of
oil as far as it is incidental to normal navigation. This situation is not

covered by the term 'dumping', as it is used here.

Art. 25(1) of the Convention is dedicated to the sinking of radio-active wastes. According to this provision, the contracting States are under the obligation to avoid the contamination of the oceans by radioactive wastes. An analogy with other, possibly as dangerous, matters is not permissible due to the exceptional nature of this provision. It could, however, be argued that, since 1958, the disposal of non-radioactive wastes has increased considerably, and that, therefore, it would be in keeping with the purpose and the objective of the Convention to include also non-radioactive, but highly dangerous, matter. This is countered by the clear wording, as well as by an *argumentum e contrario* from Art. 25(2), where with regard to 'other harmful agents', only the obligation for co-operation of the parties to the treaty with the 'responsible international organizations' is laid down. Thus, the High Seas Convention regulates the question of dumping only to a very limited extent.

Beyond this, the freedom of the seas, which is recognised by international treaties and customary international law, has to be the basis for consideration with regard to the high seas. In Art. 2 of the Convention on the High Seas, some aspects of the freedom of the seas are listed, of which the 'licence to pollute' (Le Gault)[4] is not a part. The disposal of waste is only one aspect of the freedom of use as generally accepted under international customary law (which means that it is incorporated into Art. 2 of the Convention on the High Seas). So, proof of a prohibition of dumping would have to be shown by analysing State practice.

1.2

The territorial sea is subject to the territorial sovereignty of the State. It can, therefore, permit its nationals to dispose of waste, as long as this does not affect bordering territorial seas or inhibit the innocent passage of foreign ships. The coastal State is also authorised to prohibit dumping by foreign ships in territorial waters, in so far as those are not part of the normal operations of the vessel and cannot therefore be included in the right to innocent passage.

1.3

The contiguous, exclusive economic, fishery and security zones are not subject to territorial sovereignty, but are part of the high seas.[5] The authority under international law to declare such zones does not include the competence to set rules which make possible an action against

foreign vessels which dump wastes. In connection with the discussion
of an expansion of territorial sovereignty, i.e. the sovereignty of use
beyond the territorial waters, one has often employed formulations
which suggest a competence with regard to the protection of the marine
environment.[6] In fishery and exclusive economic zones some States have
made regulations on environmental protection with regard to fisheries
and concerning environmental dangers derived from navigation. A
corresponding authorisation of the coastal States, however, has not yet
been accepted in the legal practice of States with the necessary clarity
and uniformity.

In 1970, Canada declared a special environmental protection zone in
the Arctic waters.[7] The Act prohibits, in Art. 4, the dumping of harmful wastes. The Canadian example has found nearly no followers in the
international practice of today. Without prejudice to its value for
environmental policy, it can be stated that it has not given any decisive
impulse to the development of legal regulations with regard to the
dumping of wastes at sea.

1.4

The coastal State's control over its continental shelf, as recognised by
customary law, has no implications for the dumping of waste. Continental shelves are not part of the territory of the coastal State. It has,
however, the competence to regulate the exploitation and investigation
of the continental shelf, and to exclude other States from it: thus, the
competence for regulation includes the dumping of waste in connection
with the exploitation and investigation of the shelf, notably such from
fixed platforms. Art. 5(2) of the Convention on the Continental Shelf
empowers the coastal State to establish security zones around platforms
and man-made structures, within which they are to take measures for
the protection of living marine resources from harmful influences.[8] The
competence for regulation does not include foreign waste-dumpers in
general, even if the wastes are finally deposited on the continental shelf
or have other effects on it, such as being harmful to the 'living organisms
belonging to sedentary species', as quoted in the Convention on the
Continental Shelf in Art. 2(4). For the area of the epi-continental sea
the same rules are valid as for the high seas in this respect.

1.5

Furthermore, the legal source of general principles of law has to be
considered, in particular the prohibition of the abuse of rights and the
principle of 'good neighbourhood'. It is extremely doubtful whether a

prohibition of legal abuse can be based on a general principle of law. This would mean a prohibition against exercising owner's own rights in a manner harmful to others.[9] Just as doubtful is the question whether, and to what extent, this would include the prohibition of the pollution of the high seas.[10] Speculations on the value of prohibition of legal abuse can become merely academic in particular cases, where there is no harm to subjects of international law at issue, but where nevertheless the ecological balance of the sea is adversely affected, which means that at best the 'community of international law' is concerned.

Also, the principle of 'good neighbourhood', which has so far only found partial expression as a principle in international law, is not very helpful in this context – due to its vagueness, it can only possibly serve as an aid of interpretation for other norms, but does on its own not contain any tangible basis for prohibition.[11]

The survey of the legal situation before the Conventions entered into force can herewith be concluded: in all respects, it yields only unsatisfactory regulations.

2 THE CONVENTIONS

2.1 The Oslo Convention

The Convention on the Prevention of Marine Pollution by Dumping from Ships and Aircraft was signed on 15 February 1972 by the representatives of twelve States. According to Art. 23, the Convention entered into force on 7 April 1974. According to Art. 22 of the Convention, further States can be invited to accede. The Convention is divided into 27 articles and three annexes. Arts. 1 to 15 and 19 contain regulations of substantive law, Arts. 16 to 18 regulations on the Commission that is to be created for the implementation of the Convention, and Arts. 20 to 27 constitute the final clauses.

Art. 19 of the Convention contains the legal definition of the terms 'dumping' and 'immersion', as well as that of 'ships and aircraft' (French: 'navires et aeronefs'). According to this definition, 'dumping' presupposes intent, but not an intent with regard to the harmfulness of the operation, but only with regard to the purpose of the operation. Excluded is immersion that is incidental to the 'normal' operations of vessels or their equipment. Dumping is not the issue when matter is disposed of for a purpose other than its removal. This is not dependent on the subjective intentions of the disposer, but hinges on objective purposes of the action. The problems that this limitation of the term 'dumping' create are solved by the counter exception; 'immersion' has to be considered as dumping, if another result were contrary to the objective of the Convention. Until now, this provision has not gained any practical importance. Burning of wastes at sea is not included in the definition of the term 'dumping'.

2.1.1 Objective

In its first four sections the Preamble of the Oslo Convention is put in the overall context of political and legal efforts for the conservation of the marine environment. A preamble is an important instrument for the interpretation of treaties. This has been recognised in Art. 31(2) of the Vienna Convention on the Law of Treaties of 23 May 1969.

In an emotional style that is usually reserved for the high seas and the subsoil of the sea, the Preamble of the Oslo Convention points out the aspect of the common weal: 'for all peoples' the living resources of the sea are of vital importance.

With regard to the different types of pollution, it is deemed important to take measures for prevention, such as to develop products and processes for a reduction of the amount of waste. The problem stated here is of considerable practical and political significance, for a demand to develop new products that will reduce the amount of waste would directly affect the economic structure of the signatory States, which have so far hardly taken any measures to limit the so-called integral wastage.

The qualified wording 'should be developed', and its placement in the Preamble nearly excludes the possibility that States might make mutual complaints on this point under the Oslo Convention, although mere declarations of intent are not without consequences in international law.

Further on, the special responsibility of the neighbouring States to the North Atlantic for the protection of the waters of this area is stressed, in a clause that shows the tendency to submit parts of the high seas with regard to certain matters to the regime of the neighbouring States, a tendency that can also be observed in other areas of recent developments in maritime law.

Art. 1 of the Convention is a general clause that has a very wide ambit with regard to the actions to which the States pledge themselves ('all possible steps'), as well as to the enumeration of legal interests to be protected ('endangering human health, the living resources of the sea, as well as marine life, amenities and other legitimate uses'). It remains completely open whether the term 'possible' refers only to the respective up-to-date standard for the prevention of marine pollution, or whether

it refers also to economic justifiability, or even to the limitations on
its implementation in domestic policy. The opaque wording will make it
difficult to reproach States that pursue a more reluctant policy that
might result in a violation of Art. 1. An interpretation that sees Art. 1
of the Oslo Convention in the perspective of the programmatic phrases
in the Preamble does result in the conclusion that not just any action
of a State to fight against environmental pollution through ocean dumping is sufficient, but that a serious striving for effective measures is
imperative.

Art. 2 defines the purview of the Convention. This is

> the high seas and the territorial sea which are situated (a) within
> those parts of the Atlantic and Arctic Oceans and their dependent
> seas, which lie north of 36° north latitude and between 42° west
> longitude and 51° east longitude

(excluding (i) the Baltic Sea and the Belts, as well as the Mediterranean
Sea) and '(b) within that part of the Atlantic Ocean north of 59° north
latitude and between 44° west longitude and 42° west longitude'. This
makes the purview identical with that in the treaty of the North East
Atlantic Fisheries Commission.

Art. 3 specifies that the individual domestic measures by which the
parties will fulfil their responsibilities under the Convention are to be
arranged in such a way that dumping of harmful matter in other parts
of the seas is prevented. The restriction on the freedom of dumping in
the Convention will cause industries concerned to move to areas for
disposal outside of the sea areas defined in Art. 2, in so far as this will
still be profitable. Due to its wording, Art. 3 could be interpreted in
such a way that States are obliged under international law to prevent
this. Such an interpretation is countered by the argument that the same
result could have been reached by expanding Art. 2, or avoiding the
exact demarcation of the sea area. It seems adequate to see Art. 3
merely as a programmatical request to work against a transfer of dumping to sea areas further away, which can also mean the voluntary
expansion of the measures postulated in the Convention to other areas
of the sea.

Art. 4 demands a harmonisation of national policies of the signatory
parties; they are to enter into permanent bi- and multilateral exchanges
of information.

2.1.2 The Regulations

The concrete regulations of the Oslo Convention start with Art. 5. According to this article, the dumping of matter listed in Annex I is prohibited. The Annex forms part of the text of the Convention, without this being stated explicitly (cf. Art. 31(2) Vienna Convention on the Law of Treaties). A number of particularly dangerous substances (and substances which are not, or only very slowly, rendered harmless) are named: especially organosilicon and organohalogen compounds, mercury, cadmium, certain synthetic materials, as well as carcinogens. Art. 5 is formulated like a prohibitive rule of national law. It imposes the responsibility on the States to prohibit the dumping of the substances concerned and to implement this prohibition, in as far as this is possible by means of personal and territorial jurisdiction. The rigid wording does not mean that the States would give a guarantee of this kind: should there be dumping of matter listed in Annex I, responsibility according to international law is only incurred if the criteria of accountability under the international law of state responsibility are met.[12] Accordingly, the dumping of the substances by government agencies does in any case result in the liability of the State concerned. Thus, the competence under constitutional law for the disposal of waste is irrelevant. The State can also be held responsible when dumping is authorised by a permit, the granting of which is contrary to the terms of the Oslo Convention. The State can only be made accountable for the actions of private citizens if it can be reproached for the particular action.[13] Apart from the case of granting a permit contrary to the responsibilities under international law, as mentioned above, one has to consider here the cases of wrong advice given about the procedure for the granting of permits (e.g. a government agency's information that certain dumping does not require a permit) and negligence in monitoring.

Art. 6 is concerned with a second group of wastes: in particular, arsenic, lead, copper, zinc are mentioned, as well as cyanides, fluorides and insecticides; scrap metal and bulky wastes; non-toxic substances that can have harmful effects in quantity. Art. 6 specifies the obligation for granting permission in each case, which only applies if substances are to be dumped in a certain quantity that is to be determined by the Commission, or when they are contained in other wastes. The term 'each case' ('dans chaque cas') is not defined in more detail in Art. 6. In view of the many different methods employed in disposing of waste at sea, this is regrettable. Theoretically, one possibility would be to make each single run for the purpose of dumping subject to permission. This does not seem appropriate, however, when several vessels are needed to transport wastes coming from the same source. Should one consider one permit as sufficient in those cases, it seems obvious that one should also be more liberal with regard to time, should the disposal of wastes from one source be necessary in specific intervals.

Art. 6, sentence 2, furnishes guidelines for the assessment of decisions by referring to the Annexes II and III, which contain reservations with regard to certain substances and objects.

Art. 7 comprises all substances and objects not covered by Arts. 5 and 6, and makes their dumping dependent on an official approval ('agrément'). The selection of a different term suggests that here no permit is needed in each case, and therefore the issuing of general permits is permissible. Here one could also think of an unlimited licensing subject to revocation. A legal exception to the basic prohibition of dumping is only compatible with Art. 7 if the act names the factual preconditions for dumping not subject to permission, and these are covered by the regulations in the Convention (cf. also Arts. 8, 9 and 19(1)). The official criteria for the exercise of the discretion to permit dumping are again specified by reference to Annex III.

2.1.3 Emergencies

Art. 8(1) declares that, in emergencies, no responsibility under international law for a violation of Arts. 5, 6 or 7 arises. One should not be tempted to interpret Art. 8(1) extensively, since it also terms the security of the vessel employed as an interest to be protected (which means that a material risk that can be countered by an appropriate insurance is listed) as separate from the danger for human lives. On the contrary, the objective of the Convention and the systematic reiteration of the fundamental prohibition on dumping makes high demands on those alleging an emergency situation. Furthermore, the 'vital role' of a sound ecology, as stated in the Preamble, makes a differentiation of the matters to be dumped expedient; and with regard to Art. 8(1), one has also to refer to the principle of relativity that is present in all international emergency law, for example if in the case of a threat merely to the security of the vessel, a large amount of mercury had to be dumped into the sea in order to save it, that would not be sufficient.

In any case of emergency dumpings there is an obligation in Art. 8(1) to report those dumpings to the Commission. Art. 8(2) excludes from the prohibition against any dumping in Art. 5 substances which occur only as 'trace contaminants' in waste, provided they have not been deliberately added 'for the purpose of dumping'. The remarkable feature of this provision is the fact that for the malevolent intent of the (as a rule) private disposer, the State is made accountable under international law, although the addition takes place at a time when the State does not have the possibility of exerting any influence. This constitutes a real guarantee against dumping, which is in keeping with the tendency to eliminate the element of tort in the liability of States in the case of dangerous procedures. The significance of the norm is, however, only minor, since it will hardly be possible for a party to the treaty to furnish the corresponding evidence.

It is provided in Art. 9 that in an emergency situation, where a substance cannot be disposed of on land 'without unacceptable danger or damage', the State must notify the Commission, which may recommend a suitable method of disposal.

The application of the term 'recommendation', which is employed quite often in international organisations,[14] makes it obvious that the recommendations of the Commission as to disposal are not binding. Further, the obligation on States to report the method of disposal ultimately used (as stated in Art. 9), leads to the same conclusion.[15] This obligation to report is not only relevant as a mere obligation of 'report on execution', but rather as an obligation to give a detailed account on how and how far the recommendation was followed, which can in an individual case also signify deviations from the recommendation. In spite of this, the recommendations of the Commission are by no means without legal significance. The Convention aims at comprehensive co-operation, especially through the work of the Commission, and demands that States 'seriously' consider the recommendations, i.e. under strict observance of the principle of loyalty and good faith, and not to deviate from it at will and without good reason. This interpretation is also supported by the fact that in sentence 4 all States are obliged to assist one another in such situations, which makes the commitment more tolerable for a single State.

It also has to be considered whether the occurrence of an emergency was avoidable for the State concerned, for example by means of long-range comprehensive planning of waste disposal, development of necessary procedures or their promotion, etc. (an obligation that is also mentioned in the Preamble and in Arts. 1 and 12). Should Art. 9 serve as a justification for a State which, on this issue, is reluctant to realise the objectives of the Convention, its success would be impaired.

Art. 10 provides that, prior to their decision, the responsible authorities have to ascertain in each case the composition of the waste, according to the criteria of Annex III. The term 'to be ascertained' ('s'assurer') does not imply that the official agencies have to carry out investigations on their own. It is permissible to have the investigation carried out by the applicant or the producer of the waste, provided that the authorising agency obtains information with regard to the specific composition of the waste.

2.1.4 Enforcement

Art. 11 requires States to file all records on dumpings and to furnish these to the Commission, and also creates the preconditions for the harmonisation of authorisation practice, as provided for in Art. 4.

Art. 12 requires States to establish complementary or joint scientific and technological research programmes, including research into other methods of disposal of harmful agents. Art. 12 contains also an obligation to exchange information. This obligation is formulated in the authenticated texts without any restrictions. Therefore, any results gained in such research projects would have to be made available to the parties. This does not only refer to the results reached jointly, but also to the research work done nationally, in so far as it is carried out on behalf of the State. The Convention does not consider the problems resulting from research projects that were merely stimulated by State aid, but remained under private direction.

Art. 15 determines who is to be affected by the regulations under Arts. 5, 6 and 7 that have to be given effect to by the State. The nation-ality of the persons concerned is of no significance. The determining factors are the registration of the ships and aircraft (Art. 15(1)(a)), the location of loading (Art. 15(1)(b)) and the territorial waters through which dumping vessels may pass (Art. 15(1)(c)).

Sub-paras. (a) and (b) make possible a clear demarcation that will hardly leave room for any doubts. Problems could possibly arise only with regard to the subjective component of 'for the purpose of dumping' in sub-para. (b); should the authorities of the State, where a load is taken on, be deceived about the purpose of the voyage (which means that dumping is intended to be done secretly), then the deceived State is responsible under international law, following the wording of Art. 15(1)(b), since the 'purpose' of dumping already existed upon loading. This result may be unfair in some individual cases. It is preferable to have an interpretation that leads to responsibility only if it was possible for the State concerned to recognise the true situation (whereby high standards should be set regarding the obligation to monitor). The simultaneous responsibility of several States is possible after Art. 15. Paras. (1)(a) and (b) are without legal consequences under international law; it is, however, conceivable that States will decide to follow the decision reached by another State, without investigation of their own. Sub-para. (c) takes into consideration the intentions of the disposer of the waste, while the subjective facts are made objective by this formula: does the agency suppose that dumping in its territorial waters is intended? Since the territorial waters are under the sovereignty of the coastal State, the State is in any case entitled to prevent dumping in this area, but the Convention creates a corresponding obligation. Art. 15(1)(c) merely contains the obligation to apply the existing competences under international law. Actions on the high seas, or even against vessels in foreign territorial waters, are therefore not permissible, even though this might be a means to implement the prohibition of dumping in one's own territorial waters. In as far as this involves waters of contracting parties, they should be bound according to the import of the Convention to assist the State which is obliged in terms of sub-para. (c) to fulfil its duties as far as may be reasonable.

The following paragraphs of Art. 15 serve to concretise para. (1). In this way para. 2 imposes a duty to monitor the circumstances that give rise to the suspicion of dumping without permit. It leaves it to the State to decide on the appropriateness of reporting on these measures. The aim of the treaty, as laid down in the Preamble, requires that the 'appropriateness' ('opportun') of reporting be not unduly limited, and requires States to interpret this vague term in conformity to the objectives of the treaty, which means that it shall not suffice merely to protect the repute of national industry.

Each of the contracting parties has to ensure the prevention of contraventions of the treaty.

Art. 15(5) requires States to co-operate in the development of procedures for the effective application of the agreement, 'particularly on the high seas'. This point focuses on the difficult question of the enforcing rules on the high seas over ships flying a foreign flag. Although the contracting parties do not provide for a dispensation from the prohibition against enforcing rules against foreign vessels (under customary law), they reserve for themselves the possibility of modifying existing international law which would, of course, be valid only for the flag carriers of the contracting parties.

2.1.5 The Commission

Members of the Commission, established under Art. 16(1), are representatives of all the contracting parties. Following Rule 10, No. 1 of the current Rules of Procedure, a meeting takes place at least once a year.

Art. 17 enumerates the tasks of the Commission and refers in para. (e) to 'all other' tasks necessary according to the Convention. Of special importance is the task to work out recommendations for an adaptation of the Annexes to the latest standards in technology and to economic changes (Art. 17(d)); these recommendations have to be adopted unanimously, and are subject to the agreement of the governments (Art. 18(2)).

This means that the Commission is not in the position to change the Annexes unilaterally. Art. 17(c) bestows on the Commission the task of monitoring the condition of the seas, the effects of measures taken and the feasibility of improved measures. This task can only be fulfilled by the Commission if it is vested with the necessary administrative and scientific resources, which are not mentioned in the agreement. The task as stated in (c) will, therefore, have to be recognised as fulfilled when an appropriate evaluation of the data gathered through the activities of the individual States has taken place.

Whether or not the remainder of the contracting parties are entitled to change the annexes among themselves, in the absence of agreement of one contracting party, will depend on the individual case. This results from Art. 41(1) of the Vienna Convention on the Law of Treaties: a corresponding procedure is not provided for in the Oslo Convention (cf. Art. 41(1)(a) VCLT); at the same time there is no explicit prohibition (cf. ibid., (b)). That a non-regulation does not equal a 'prohibition' is evident from the separate statement of these two cases. A revision *inter se* is therefore permissible if it does not inhibit the overall objective of the treaty.

2.2 The London Convention

The Convention on the Prevention of Marine Pollution by Dumping of Wastes and Other Matter was signed in London on 29 December 1972, about ten months after the Oslo Convention.[16]

The definition of the term 'dumping' in Art. III of the London Convention is divided — as in Art. 19 of the Oslo Convention — into a positive and a negative description. The basis is identical (in spite of a difference in wording). Art. III(1)(a)(ii) mentions, in addition, the disposal at sea *of* vessels, aircraft, platforms or other man-made structures that are not covered in the Oslo Convention, since it regulates only the disposal *from* vessels and man-made structures.

As in Art. 19(1)(a) of the Oslo Convention, though in more detail, Art. III(1)(b)(i) of the London Convention determines that the disposal of wastes incidental to or derived from the normal operations of vessels or man-made structures and their equipment does not qualify as dumping.

The definition in the London Convention includes also the burning of wastes at sea. This is not expressly incorporated, but Art. III(1)(b)(i) includes the disposal of wastes or other matter transported by or to vessels, aircraft, platforms or other man-made structures at sea, operating for the purpose of disposal of such matter or derived from the treatment of such wastes or other matter. This somewhat awkward description includes the residue resulting from the burning of wastes.[17] Accordingly, the contracting parties established an *ad hoc* working group to clarify the question of the burning of wastes at sea.

The placement of matter for a purpose other than disposal does not qualify as dumping.

2.2.1 Objective

At the beginning of the London Convention stands the general pledge to fight marine pollution by dumping of matter (Art. I). The interests which are to be protected from the effects of hazardous matter are identical to those mentioned in Art. I of the Oslo Convention, whereas the extent of measures required in the English text (although not in the French text) is more limited: instead of 'all possible steps' merely 'all practicable steps' are called for. In practice this difference will be insignificant.

Art. II of the London Convention notes that States are only obliged to fulfil the Convention 'according to their scientific, technical, and economic capabilities'. This reservation implies — taken literally — a considerable restriction. Apart from the cases of *de facto* impossibility, a State, attempting to disengage itself from the obligations resulting from the Convention by reference to Art. II, can only be bound by the limitations of loyalty and good faith. This may lead to a restrictive interpretation of the norms, particularly in cases where the economic feasibility of an obligation is negated, since the recognition of the value of the protected legal interest ('marine environment' in the first paragraph of the Preamble) demands considerable economic efforts.

2.2.2 The Regulations

Art. IV(1) and (2) contain corresponding regulations to Arts. 5, 6 and 7 of the Oslo Convention. In particular, the threefold division is maintained:

   (i)   generally prohibited matter (Art. IV(1)(a) in connection with Annex I(12)(a));

  (ii)   matter, the dumping of which requires a prior special permit

(Art. IV(1)(b) in connection with Annex II);
(iii)   and matter the dumping of which requires a permit, but only
a general one (Art. IV(1)(c)).

Art. V(1) of the London Convention, dealing with exceptions to Art.
III, is identical in content to Art. 8 of the Oslo Convention, but is
worded more carefully and in more detail; it does not depend on the
judgement of the contracting party concerned, but is worded 'objectively',
and specific restrictions have been included in the text ('real threat', 'the
only way', 'relativity', 'obligation to minimize damage'). As in Art. 8(1)
of the Oslo Convention, there exists the obligation to report (Sub-para.
2).

The remainder of the exceptions to the prohibition on the dumping
of matter differ textually in the Conventions, but are similar in content.
Art. 9(2) of the Oslo Convention provides that the prohibition is not
valid when the matters concerned are only present as trace contaminants,
but that in this case the obligation to acquire a permit exists, corresponding to the provision of Annex I(No. 9) of the London Convention. The
lack of a qualification (cf. the relevant clause in Art. 8(2) of the Oslo
Convention) stating that the exception does not apply when the matter
is added to the waste for the purpose of dumping, can be explained by
the general principle of the prohibition of legal abuse.

A special exceptional regulation for the substances listed in Annex I
can be found in Art. V(2) of the London Convention, which is comparable
to Art. 9 of the Oslo Convention; the English and French texts employ—
the same as there—the adjectives 'unacceptable' ('inacceptable') to
describe the necessary situation of risk, and show only minor differences
on the whole. The procedure to be followed in such emergency situations
is regulated differently in each Convention. Whereas Art. 9 of the Oslo
Convention requires only a consultation with the Commission, Art.
V(2) of the London Convention demands, beyond this, consultations
with all countries ('pays') possibly concerned. While the text of the
Convention always employs the term 'contracting parties', this regulation
makes it clear third countries are included, though they have no right to
consultation. IMCO itself has to engage in further consultation before it
issues a 'recommendation'. How far the recommendation is legally
binding is set forth in more detail than in Art. 9(3) of the Oslo Convention;
the result is a similar commitment according to discretion; the criteria
are the 'times available', consideration of the 'general obligation to avoid
damage to the marine environment' and 'reasonable' consideration.

2.2.3 Enforcement

Those concerned by the measures taken by national authorities for the
implementation of the Convention have been defined in more detail in
the London Convention than in the Oslo Convention. Art. VI(2) of the
London Convention provides that the responsibility for issuing a permit
is determined by the location of loading or registration of the loading
vessel. Art. VII(1) provides that the measures required to implement the
present Convention are to be applied to those vessels and aircraft under
the jurisdiction of the contracting States (cf. Art. 15 Oslo Convention).

Art. VII(2) of the London Convention is identical in wording to Art.
15(3) of the Oslo Convention: each party takes in its territory the
appropriate measures to prevent and punish conduct in contravention
of the Convention. In areas of the seas not within the territory of the
contracting parties, there exists no such authority. But, at least, the
contracting parties are pledged to work out procedures for the reporting
of vessels from which dumping takes place in contravention of the
provisions of the Convention. This has, however, so far not been put
into practice.

The exclusion of vessels with sovereign immunity (Art. VII(4) of the
London Convention) is a reference to the diverging opinions on this
point, due to the participation of socialist States in the Convention. This
is made more acceptable by means of Art. VII(4), whereby States are
obliged to ensure that vessels with sovereign immunity act in a manner
consistent with the object and purpose of the Convention. The exclusion,
therefore, only prevents the execution of foreign jurisdiction over such
vessels, but does not establish a licence to dump in contravention of the
Convention.

Art. IX of the London Convention accounts for the technological
and institutional differences of the contracting parties, which is a reaction
to a problem that does not arise within the homogeneous circle of the
Oslo Convention. States that so far are less capable of combating the
consequences of dumping of waste at sea, particularly with regard to
monitoring and treatment of wastes, shall profit from the 'know-how'
of the more advanced States by means of co-operation. Aid is to be
granted among the countries concerned. This provision is part of the
general efforts for technology transfer. In practice, such requests reach
the departments for technological co-operation and for marine environment of IMCO via their regional consultants. The department for marine
environment operates either on the basis of its own knowledge, or in co-
operation with scientific institutions, the cost of which is borne, under
certain conditions, by IMCO or the United Nations.

Pursuant to Art. X of the London Convention, States have to develop procedures for the assessment of liability in the event of damage through dumping and for the settlement of disputes. This work has not yet been concluded. The provision cannot be called a *pactum de contrahendo*, since it is merely an obligation to develop such regulations ('entreprendre l'élaboration'), and not to give these a legally binding character (which could be done by means of alterations of, or amendments to, the treaty, or through the conclusion of a separate treaty). Therefore, declarations about the intention to apply the eventual rules would also satisfy Art. X.

The obligation in Art. XI (consideration of procedures for the settlement of disputes) has so far not been fulfilled either.

Art. XII encourages international co-operation with the aim of taking additional, individually listed measures for the protection of the marine environment.

Art. XIII of the London Convention makes clear that its adoption shall be without prejudice to a codification of the international law of the sea by the Third United Nations Conference on the Law of the Sea. This is why a consultative meeting on the rights and responsibilities of a coastal State for the application of the Convention in a zone adjacent to its coast ('dans une zone adjacente à ses côtes') was postponed until after the conclusion of the Law of the Sea Conference.

Art. XIV lays down, in detail, the tasks bestowed by the contracting parties on IMCO, in particular those referring to consultation, co-ordination and application of the Convention. Revisions of the Convention are decided on by a two-thirds majority of the contracting parties represented (Art. XIV(4)(a) and Art. XV(1)). A revision enters into force only for those parties that adopted it. Procedures and time limits are extensively set forth in Art. XV.

The Joint Group of Experts on the Scientific Aspects of Marine Pollution (GESAMP) co-operates with the contracting parties and IMCO, as well as in consultations in scientific and technical questions in connection with the Convention (Art. XIV(4)(b)). This is a scientific working group of several of the Specialised Agencies of the United Nations and of the International Atomic Energy Agency, whose work is headed by an administrative secretary provided by IMCO. Further *ad hoc* working groups, staffed by the contracting parties, dealing with scientific and legal questions of dumping and burning at sea, have been established. Subsidiary bodies of IMCO, such as the Sub-Committee on Safety of Navigation and the Sub-Committee on Bulk Chemicals (both subgroups of the Marine Safety Committee), have in special cases dealt with dumping.

2.3 The Helsinki Convention

The Convention on the Protection of the Marine Environment of the
Baltic Sea Area (the Helsinki Convention) was signed on 22 March
1974 by seven neighbouring States to the Baltic Sea. For the first time
in the history of international law this Convention endeavours to
regulate the protection of one sea area from *any* kind of pollution. The
Baltic Sea, being a shallow marginal sea, is especially vulnerable to
pollution and is, in addition, extensively used by industry, navigation
and tourism.

The provisions of the Helsinki Convention have been discussed
intensively in the literature;[18] what follows are only those regulations
that are of special relevance to the dumping of waste.

Art. 2(3) contains the definition of the term 'dumping', which is
identical to Art. III(1) of the London Convention, apart from one
stylistic variation. Burning at sea is, therefore, also included in the
Helsinki Convention.

Art. 3(1) lays down norms for a comprehensive obligation (through
legislative, administrative and other measures) to protect the Baltic Sea
from pollution, which according to para. 2 shall not lead to a pollution
of other areas of the sea.

The problem of dumping is treated in relative detail in Art. 9. The
contracting parties are obliged to introduce in national law a general
prohibition (Art. 9(i)). Only dredged spoils may be dumped, subject
to a prior permit by the authority responsible in accordance with the
provisions of Annex V, Rule 1, of the Convention.

Like Art. V(2) of the London Convention, and differing from Art.
9(1) of the Oslo Convention, the emergency provisions in Art. 9(4) of
the Helsinki Convention do not leave the definition of the emergency
to the discretion of the State seeking to be relieved from the prohibition.
The factual preconditions are more demanding than in the parallel norm
cited: with regard to the vessel, complete destruction or total loss must
be impending; every probability must indicate that the damage through
dumping will be less than the damage otherwise incurred.

In the application of Art. 9(4), following Art. 9(5), the provisions of
Annex VI have to be applied, including extensive obligations for co-
operation and information. In addition, the Commission for the Protection
of the Marine Environment of the Baltic Sea, to be established under
Art. 12 of the Helsinki Convention, has to be reported to (Rule 4 of
Annex V contains an exact catalogue of the information to be reported).

The co-operation of the signatory parties (cf. here Art. 11 and Annex
IV) is institutionalised by the Helsinki Commission. Its tasks are regulated

in Art. 13. Significant among them is the obligation to observe the implementation of the Convention, keep the contents up to date, define criteria for the monitoring of pollution and aim for a minimisation of pollution.

The procedures for revision of the annexes are noteworthy. In contrast to the two Conventions on dumping, a contracting-out procedure (Art. 24) was chosen. The Commission can present proposals for revision to the contracting parties and set a time limit within which objections have to be articulated, otherwise the provision becomes binding. It can be concluded from Art. 24(a) that also a revision *inter pares* is possible.

3 THE THIRD UNITED NATIONS CONFERENCE ON THE LAW OF THE SEA (UNCLOS III)

UNCLOS III deals (in its Third Main Committee) with questions of environmental protection, together with the areas of 'research' and 'transfer of technology'. The Informal Composite Negotiating Text (ICNT) deals in Part XII ('Protection and Preservation of the Marine Environment') explicitly with the question of dumping of waste at sea.

Article 1(5) of the ICNT contains a legal definition of the term 'dumping' – which is literally identical with that of Art. III(1) of the London Convention, apart from an explicit inclusion of burning (Art. 1(5)(a)). Earlier drafts had only provided for a reference to the London Convention.

Article 195(3)(a)(iii) of the ICNT lists harmful substances (particularly such that do not disintegrate), as an area of special emphasis, where measures have to be taken by the contracting parties. This makes the general statements about a future marine environmental policy also relevant for the field of waste dumping, for example avoiding replacing a harmful disposal method by another or mere exchange of the disposal areas (Art. 196), obligation for co-operation on a global and regional level (Art. 198), obligation to report imminent damage (Art. 199), promotion of scientific co-operation (Art. 201 ff.), scientific and technical assistance for development countries (Art. 203ff.) and entertaining a common monitoring system (Art. 205ff.).

The duties of States in relation to the dumping of waste are regulated in Art. 211 of the ICNT. Accordingly, national *legal* provisions have to be created which reduce and control dumping (para. 1), and States must take *'other measures'* to 'prevent', 'reduce' and 'control' dumping (para. 2). From the division into 'legislative' and 'other measures' it can be concluded that, apart from legal control, political action to prevent pollution by the dumping of wastes is also required. The national laws,

provisions and measures shall ensure that no dumping without permission
from the responsible authority will take place. Art. 211(4) of the ICNT
specifies the general obligations for co-operation that are determined by
Arts. 201 and 202 of the ICNT.

An important regulation, at least partially diverging from international
law, is to be found in Art. 211(5) of the ICNT, following which the
dumping of waste is only possible in foreign territorial waters and
exclusive economic zones, and on foreign continental shelves, with the
express prior consent of the coastal State. With regard to territorial waters,
the regulation of Art. 211(5) of the ICNT is only novel with respect to
the introduction of the obligation to consult; other than that dumping
without consent is not permissible. Concerning the exclusive economic
zone and the continental shelf, the authority to control foreign dumping activities goes beyond that under existing law. In this regard, Art.
211(5) corresponds to Art. 56(1)(b)(ii) of the ICNT, which confers on
the coastal State jurisdiction for environmental protection in the exclusive
economic zone. There is, however, no point of reference in the regulations
concerning the continental shelf (cf. Art. 76 ff of the ICNT). This results
in the strange situation that the negotiating text only provides pollution
jurisdiction in the area of the continental shelf with regard to dumping
of waste. This discrepancy will remain, without major practical consequences, due to the fact that there are only very few continental
shelves wider than 200 nautical miles (the extent of the EEZ).

Since the high seas will be extensively divided up into zones by the
convention planned, the national permission procedures will have to
operate according to Art. 211(4), under constant observation of the
possible rights of other States, where the already existing institutions,
particularly on a regional level (as the Oslo and Baltic Sea Commissions),
can serve as the adequate framework to formalise and organise the
granting of permissions.

Art. 217(2) makes the obligations under para. 1 subject to the cancel-
ling condition that another State has taken such measures; if it has
commenced measures according to para. 1, the obligation is annulled. It
seems doubtful whether this regulation, considering merely a sequence
in time, without defining the commencement of the measures more
concretely, is appropriate; or whether it does not allow the possibility
for a good excuse for one's own inactivity. In any case, one will be able
to see a reference to Art. 217(2) as an abuse when insufficient measures
were taken by the other State and adequate information was available.

Art. 238(1) of the ICNT makes clear that international treaties
concluded prior to or after the proposed convention, containing more

specific obligations with regard to marine environmental protection, remain unaffected. According to para. 2, they shall be interpreted in such a way as to be compatible with the principles and objectives of the convention. This regulation is, of course, only applicable to the circle of contracting parties.

EDITORIAL NOTE

The UK Government has given internal effect to both the Oslo and London Conventions in the Dumping at Sea Act 1974. Prior to the Act, the only controls over dumping at sea were voluntary arrangements, but the Act now regulates 'dumping'[19] within the UK territorial waters[20] and dumping by UK vessels or aircraft (including military vessels and aircraft) anywhere in the world.[21]

Dumping is prohibited[22] unless a licence is obtained from the Ministry of Agriculture, Fisheries and Food[23] (but licences granted by the appropriate authority in countries which are parties to the Conventions may be recognised[24]). In deciding whether or not to grant a licence, the authority must consider the effect which the proposed dumping will have on the environment and, if a licence is granted, conditions may be imposed for its protection.[25] Breach of these conditions will result in revocation of the licence,[26] and any person who knowingly practises deception in order to obtain a licence commits an offence.[27]

Unauthorised dumping may result in a fine or a term of imprisonment or both,[28] but there are various defences open to anyone charged with contravening the Act. If such a person can show that the dumping was done to secure the safety of the vessel or aircraft, etc., from which dumping took place, or to save life, and that he took steps within a reasonable time to inform the licensing authority of his action, he will escape liability unless, in the opinion of the court, the dumping was unnecessary and unreasonable.[29] It is also a defence that the person charged was acting on instructions from his employer or on information supplied by others and that he had no reason to suspect that the information was false or misleading, provided, in both instances, he took reasonable steps to ensure that he was not committing an offence.[30] If the dumping takes place outside UK territorial waters, but from a British ship, aircraft or hovercraft, it is a defence to establish the loading took place in a Convention State and was authorised by that State.[31]

The task of enforcing the provisions of the Act lies with a 'British enforcement officer',[32] who may inspect land (including private dwellings unless they are used for business purposes), vehicles, aircraft and hovercraft in the UK, ships in UK ports and UK vessels, etc., wherever they may be.[33] During such an inspection, he may take samples, examine equipment and require the production of licences and records, of which he may take copies.[34] Any person who, without a reasonable excuse, impedes the enforcement officer shall be liable to a fine.[35]

One important point noted above is the definitions of 'dumping'. The Act provides that 'substances and articles' (a phrase which is not defined) are 'dumped' if they are permanently deposited in the sea from a vehicle, ship, aircraft, hovercraft or marine structure, or from a structure on land constructed or adapted wholly or mainly for the purpose of depositing solids in the sea.[36] The definitions in the Conventions are much narrower than this, and it remains to be seen whether the definition in the Act will create problems of construction – for example there could be some dispute about the use of the word 'solids' – but it may be that few problems will arise in that the licences are granted by the Minister from whose decision there is no appeal.

NOTES

A more detailed analysis of the subject is Ehlers and Kunig, *Abfallbeseitigung auf Hoher See* (Völkerrecht und Recht der Bundesrepublik Deutschland, 1978).

1. For details see de Yturriaga, 2 *Rev. de Instituc. Europeas* (1975), p. 49.
2. For details see Kiss, 22 *A.F.D.I.* 731 (1976).
3. See below, p. 198 (UNCLOS) and Ehlers and Kunig, *Abfallbeseitigung*, p. 96 (EC).
4. (1971) 21 *U. Tor. L.J.* 217.
5. Cf. Rojahn, 19 *G.Y.I.L.* 83 (1976).
6. Cf. Judge Padilla Nervo's Dissenting Opinion, *Fisheries Jurisdiction Case*, [1973] I.C.J. Rep. 44.
7. Text in *New Directions in the law of the Sea*, Vol. I, p. 199.
8. For details see Gündling, 37 *ZaöRV* 547 (1977).
9. See Taylor, (1972-73) 46 *B.Y.I.L.* 323.
10. Cf. Khan, (1973) 13 *I.J.I.L.* 407.
11. Ed. Klein, *Umweltschutz im völkerrechtlichen Nachbarrecht*, 1976, p. 115 et seq.
12. For details see von Münch, *Das völkerrechtliche Delikt in der modernen Entwicklung der Völkerrechtsgemeinschaft*, 1963, p. 149 et seq.; (1976) 25 *I.C.L.Q.* 509.
13. For details see von Münch, *loc. cit.*
14. Cf. Wehser, 2 *Thesaurus Acroasium* 371 (1976); Schreuer, 20 *G.Y.I.L.* 103 (1978).
15. According to Lucchini, 101 *Clunet* 776 (1974), the formulation 'en application de ses recommendations' suggests a binding force of the Commission's decisions.
16. According to its Art. XIX the Convention entered into force on 30 August 1975.
17. Different opinion: Ballenegger, *La pollution en droit international*, 1975, p. 127.

18. Cf. B. Johnson, (1976) 25 *I.C.L.Q.* 1; D.A. Boczek, (1978) 72 *A.J.I.L.* 782.

19. The term is defined in s. 1(2).

20. S. 1(1)(a).

21. S. 1(1)(b).

22. S. 1.

23. In Scotland and Wales, the licence is issued by the Secretary of State (s. 12(1)); the Transfer of Functions (Wales) (No. 1) Order 1978 S.I. 1978/282. The appropriate authority in Northern Ireland is the Department of the Environment.

24. S. 6.

25. S. 2(1).

26. S. 2(2).

27. S. 2(9).

28. S. 1(6).

29. S. 1(7).

30. S. 1(8).

31. S. 1(9).

32. S. 5(1), (2) and (3).

33. S. 5(5) and (6).

34. S. 5(8).

35. S. 7.

36. S. 1(2).

8 POLLUTION FROM LAND-BASED SOURCES

D. Alastair Bigham

Before discussing the law and practice relating to pollution of the seas from land-based sources, certain facts should be clearly stated. Although the most frequent journalistic reporting of pollution incidents relates to accidents and deliberate spills or discharges of oil which have a disastrous local effect, it should be realised that by far the greater volume of marine pollution originates in the form of discharges into fresh water which passes into the sea through river estuaries. It has been variously estimated (no doubt due to wide differences in local conditions) that between 80 and 95 per cent of marine pollution is from land-based sources.

The principal difficulty in considering this subject lies in deciding upon a means of limiting the length of discussion, bearing in mind that this book is largely concerned with international agreements and, thus, with a world-wide picture. In the case of land-based pollution, however, one is inevitably required to consider problems which are of an essentially territorial nature, and are thus governed ultimately by the detailed internal legislation of individual sovereign States, that legislation being inescapably an integral part of national law.

The approach adopted in this chapter (and in Chapter 10) is, therefore, to limit commentary to the situations in the more advanced industrialised countries, and to certain international agreements and legislation. One reason for selecting this approach is that, in the case of the developing countries, the primary industrial problem to be faced by any national government may well be that of achieving a sufficient national income to rise, or remain, above the poverty level, rather than to be concerned with immediate problems of pollution, except where the pollution is of a type which could lead to infection and epidemics. This is not to say that, in the world sense, pollution affecting the seas off impoverished countries should be ignored, but merely that, to achieve prevention, international financial aid may be required.

It is appropriate to comment at this stage that the essential reason for prevention of marine pollution relates not merely to the enjoyment of beaches by holiday-makers, but (in a very real sense) to both maintaining the world ecological balance and to meeting the increasing need to provide food for an expanding world population. This is clearly (but

approximately) illustrated by the fact that, if it is assumed that certain
broiler chickens are reared on fishmeal, it will take about 10 lb of
nettable-sized fish to produce 1 lb of chicken-meat (protein) and 100 lb
of smaller fish to produce the 10 lb of netted fish. Thus, if the resources
of the sea can be successfully husbanded, the world food supply can be
immeasurably increased, particularly if fish is used to a greater extent
in human diet.

In considering the situations in those countries having significant
industrial development, it is convenient to divide them into three
categories.

(i) In the most acute situation, certain States (as in the case of Japan)
have such a great concentration of pollutant-producing industry that
they have in some respects passed beyond the stage of having pollution
control as their immediate legislative aim to a situation in which the
provision of compensation for financial loss and for permanent and
temporary damage to health have become of principal concern.
Fortunately, this type of situation is not, as yet, widespread, but a
short commentary will be offered.
(ii) The second category to be considered will cover the majority of
countries in which manufacturing and extractive industries are a
vital element in the national economy, and which have a relatively
(but not excessively) high density of population, as in Western
Europe. The legislative provisions developed, and intended to be
further extended, in the European Community are discussed by the
present writer in Chapter 10; but the position in the UK (and principally
in England and Wales) will be referred to in the current chapter.
(iii) The third category of industrialised country is that in which,
although it is reliant upon industrial production as an essential part
of the economy, the total land area available is so great that sources
of land-based marine pollution are relatively localised — as in many
sections of the coastline of the North and South American continents,
Southern Africa and Australia. The example selected for consideration in this chapter will be the United States, since this can illustrate
certain aspects both of means of detailed control applied within large,
semi-autonomous States and of certain problems relevant to a federal
system of government.

International Co-operation

The increasingly urgent need to take positive steps to protect the world
environment has been recognised in a number of international agreements,

and notably in those reached at the United Nations Conference on the Human Environment, held in Stockholm in June 1972, which was followed, in the same year, by a summit conference of leaders of the European Economic Community, held in Paris in October, and by the Conference on the Prevention of Marine Pollution by Dumping of Wastes and other Matter, held in London in December of that year.

Subsequent meetings related to these conferences have followed and, in particular, the European Community (a title now favoured after elections to the European Parliament)[1] has pursued a complicated and intensive programme, as will be described later in Chapter 10.

It is also of great importance that widespread international agreement on environmental issues (in addition to those of security and other international co-operation) was reached at the Conference on Security and Co-operation in Europe, held in Helsinki in July 1973, continued in Geneva in July 1975 and concluded in Helsinki on 1 August 1975. This Conference resulted in Conventions signed by representatives of 35 States as to general co-operation on security, including in relation to the environment;[2] and also by six non-participating Mediterranean States, in regard to co-operation in the Mediterranean.[3]

As to environmental matters, the signatories agreed (i) to study the problems relating to the environment; (ii) to increase the effectiveness of national and international measures for its protection

by the comparison and, if appropriate, the harmonisation of methods of gathering and analyzing facts, by improving the knowledge of pollution phenomena and rational utilization of national resources, by the exchange of information, by the harmonization of definitions, and the adoption, as far as possible, of a common terminology in the field of the environment;

(iii) to take the necessary measures to bring environmental policies closer together and, where appropriate and possible, to harmonise them; and (iv) to encourage international efforts for monitoring, protecting and enhancing the environment.

The Convention then gives examples of fields of co-operation which include water pollution and fresh water utilisation; and (separately) protection of the marine environment. With regard to water pollution and fresh water utilisation, the aims are the prevention and control of water pollution, in particular of trans-boundary rivers and international lakes; techniques for the improvement of the quality of water and further developments of ways and means of industrial and municipal sewage effluent purification; methods of assessment of fresh water resources and the improvement of their utilisation, in particular by developing methods of production which are less polluting and lead to less consumption of fresh water.

As to protection of the marine environment, reference is made particularly to the Mediterranean Sea and to the London Convention on Dumping; and (in the context of this chapter) also to 'problems of maintaining marine ecological balances and food chains, in particular such problems as may arise from the exploration and exploitation of biological and mineral resources of the seas and the sea-bed'. Comment will be offered later as to the Mediterranean, under the Barcelona Convention (*post*, Chapter 10).

Further quite detailed statements of policy as to international cooperation are made, which refer to such matters as exchanges of scientists, specialists and trainees; meetings of experts; joint projects; further Conventions, etc.

The effects of these and other proposals, which have been largely accepted and put into effect in regard to Western Europe by Decisions and Directives of the Council of Europe, can be seen in some detail in Chapter 10. The application in regard to water pollution control in the United States of America is indicated in Part III of this chapter.

1 COUNTRIES HAVING A VERY HIGH LEVEL OF POLLUTION, RESULTANT IN STATE-ADMINISTERED SYSTEMS OF COMPENSATION FOR INJURY

In a number of countries where there is a high density of total population combined with very rapid, recent industrial growth, control of pollution has been extremely difficult. Considerable injury to health of sufficient severity to necessitate medical treatment has occurred, and has affected individual members of the general public. It has therefore become essential in recent years for the governments of the countries in this situation not merely to increase pollution control and control of industrial development in the planning sense, but notably to devise systems whereby the State itself takes a direct part in ensuring that the cost of treatment of individuals affected by toxic substances is met either by the industrial polluter or, in whole or in part, by the State or local government.

Civil remedies for damages are, of course, available in the courts of virtually every country, and the individual citizen suffering from such toxic effects can, no doubt, obtain compensation in due course, subject to his ability to prove causation; and subject to considerable delay and probable personal expense. It is not, however, within the pattern selected for this chapter on land-based marine pollution to discuss such private civil remedies further.

With regard specifically to land-based marine pollution, the salient features which should be stressed in relation to such extreme situations resulting from industrial pollution are:

(i) that the degree of pollution emanating from rivers and estuaries and from other land-based sources is generally far higher (with certain exceptions) than in the coastal waters of the States to be discussed later in this chapter;

(ii) that, in the short term at least, the battle for adequate pollution control has been lost in many localities in which the polluting industries responsible cannot, for national economic and local employment reasons, be closed down;

(iii) that the resort to a widespread system of payment of compensation for medical injury to the general public from toxic substances does, in a sense, make it more politically possible to continue with an unacceptably high level of industrial pollution; and

(iv) that (as will be seen below) such a system may also provide for State aid to an industrial plant causing pollution in cases where an expensive technological remedy or closure would result in local unemployment, and this form of government subsidy is incompatible with a 'free market' philosophy, in terms of commercial competition in international trade.

As an example of sophisticated thought applied to such systems of State-administered compensation, it is convenient to look at Japan. As a result of the pressing economic need to increase national income from industrial exports, due to the high density of population, Japan has concentrated on vast industrial development. In consequence, by the end of the 1960s it was evident to the State Government that much stricter land-use planning controls must be applied, but also that, the degree of toxic pollution of both the air and water having become so severe, a system to ensure payment of the cost of the medical treatment of victims of the situation must be adopted.

It is not proposed to discuss the method of pollution control applied in Japan, except to observe that the State passed the Water Pollution Control Law in December 1970, which has a bearing on land-based marine pollution, and the Law relating to the Prevention of Marine Pollution and Maritime Disaster, also in December 1970. The Government's view is that in certain areas the level of control is comparable to, and sometimes higher than, that currently effective in the European Community. The number of reports of 'red tides' (caused by titanium dioxide) has been declining. Certain Prefecture governments also apply more stringent local laws in relation to fresh water and to marine pollution.

The question of State-administered medical compensation for pollution injury is, however, of interest. Initially, in 1969, the Law for the Relief of Damage to Health provided that the cost of medical treatment caused by pollution of air and of water (but not of other types) could be met on a basis of half being payable by the polluter, and half by the public authority. However, some Prefecture governments took the view that compensation under this law was frequently inadequate, and they accordingly passed local laws which increased the responsible industrialist's liability by providing a greater range of compensation.

The measures taken by local authorities resulted, in effect, in repeal of the 1969 Law, and its re-enactment and extension in the form of the Pollution-Related Health Damage Compensation Law, December 1970. This abrogated much of the local legislation, but provided a system whereby a case of pollution causing medical damage to an individual is investigated by the public authority and, on proof of causation, the responsible polluter is required to pay the entire cost of medical treatment. It is thus the responsibility of the public authority to investigate all such cases; to assess the cost; to obtain appropriate payment from the polluter; and to transmit that payment to the medical authorities.

Although the present national legislation does not provide compensation for loss of earnings due to pollution, certain local laws continue to do so, and they may tend to illustrate, in a sense, a permissive attitude. In the Oita Prefecture there is a Local Law for the Relief of Damages to Fisheries, 1974, whereby loss of income to fishermen, caused by a 'red tide', oil spill or other water hazard, is compensated for out of a relief fund contributed to on a basis of 15 per cent from the Prefecture government, 15 per cent from municipal governments in the coastal area, and 70 per cent from private industries in the coastal area.

In the Akita Prefecture, where mining and agriculture are the major types of industry, the Prefectural government operates a system of compensation for cadmium contamination of rice crops. This requires that the farmer affected by deposits of dust reports the situation to the Agricultural Economic Association, which in turn reports to the Prefectural government. The Prefectural government then files a claim for payment with the Association of Mining, which holds funds contributed to by the mining operators. On payment, the Prefectural government transfers the money to the Agricultural Economic Association, which in turn pays the farmer, but receives the contaminated rice in exchange. However, where the polluter cannot be precisely identified, the Association of Mining pays only half the cost and the Prefectural government pays the balance; or, in some cases, the Prefectural government may pay the whole. It is therefore clear that eventual pollution of local rivers may, in fact, be a not uneconomic proposition for the mining industry.

However, in terms of the practical effect upon sources of land-based pollution, it is also significant that the Pollution-Related Health Damage Compensation Law enables the public authority to consider whether to provide financial aid to a polluter on the basis that the cost of future prevention of the particular pollution would result in local unemployment. Reference has already been made to the effect upon the market/ competitive situation; but it is, of course, significant that the Japanese Government is thus providing economic aid which will result in steps being taken to prevent escape of toxic substances into water from such specific sources. This, in itself, can serve to reduce land-based marine pollution, *provided that* such financial aid for introduction of new technological means of prevention is actively and thoroughly pursued; but it is significant that, as the writer understands the situation, the question is still viewed as a party-political issue, and the approach to problems of pollution is, as yet, by no means unified. In terms of the interests of the remainder of the world, it is of the greatest importance that pollution be stopped, rather than that it be permitted to continue by use of extensive medical compensation in order to reduce pressures of public opinion.

2 COUNTRIES HAVING THE TYPICAL PROBLEMS ARISING FROM AN ADVANCED INDUSTRIAL ECONOMY AND RELATIVELY LIMITED LAND AREA

The problems presented by the older and advanced industrial economies provide the most useful and major topic for commentary in regard to land-based marine pollution. The measures adopted or (of equal importance) envisaged are very largely also of value for consideration by those parts of the world discussed under Parts 1 and 3 of this chapter, in that they illustrate sophisticated provisions for legal, economic and technological control of pollution, which are built upon — or are additional to — well established criminal and civil remedies, and systems of national and local government. Clearly, in countries where the 'industrial revolution' occurred early (beginning more than two centuries ago), experience of environmental problems should be extensive – even if the response has not, in all cases, been satisfactory.

In considering this type of situation, a commentary is offered on the position in the UK, on the basis that *the reader should relate it to the further commentary on the European Community legislation, given in Chapter 10.* The UK is taken by way of example, since it is possibly the best illustration of effective reduction of water pollution, achieved despite a long history of high-density industrial development.

2.1 The United Kingdom: a Short Comment on the Main Provisions for Control of Water Pollution

This section is written by way of illustration and not as a precise and detailed source of reference. Thus, although the legislation referred to applies to England and Wales, it is upon the same principles as the related Scottish legislation.

As in many other European States, the earliest controls exercised over water pollution were by means of local legislation in the form of by-laws passed by the ancient cities and boroughs by virtue of powers granted under their charters. The purity of rivers has also, for many hundreds of years, been (in theory at least) often protected, or capable of protection, by the rights vested in riparian owners – i.e. owners of land on either side of a river (including ownership of the bed of the river, normally up to midstream). These rights included that the quality and, in general, volume and temperature of the water in a river must not be materially altered by other owners and occupiers of riparian land, subject to certain rights as to normal agricultural use, etc.

There were also, on occasion, specific Acts of Parliament passed in regard to localised pollution and nuisances. In addition, there have been various means of control – or at least of discouragement – of pollution, by civil and criminal actions for nuisance, which can be brought either by private individuals, or, more significantly, by the Attorney-General on behalf of the citizens in an affected locality. In the present century, there has also been made available the simple machinery of control of 'statutory nuisances' under the Public Health Acts, which largely relies upon an initiative being taken by a local authority (or, failing this, by a group of local inhabitants), the case being heard before the local magistrates. In addition, other types of statute and by-laws have long existed for the protection of salmon and freshwater fisheries, and to provide a limited capacity for control of estuarial pollution by district sea fishery committees, or earlier similar bodies, representing jointly fishery and land-based authorities; but these latter areas of statutory control are, and have been, very dependent upon initiatives by sectional interests, rather than upon public initiative. In short, as to earlier controls, damages and sanctions relating to water pollution, it can be said that these have existed in various forms for many generations, but that they have, until this century, been of limited assistance, due to the elements of delay, cost and public ignorance of legal rights. They have been further complicated by the fact that the precise type and degree of remedy has, historically, tended to be different according to whether the water itself is (a) in a 'natural' watercourse (i.e. a stream, river or natural lake fed by them; or tidal waters) or in a definable underground watercourse; or (b) percolating through underground strata; or (c) in artificial watercourses or lakes; or (d) intended for domestic use – as in a water main.

So far as discharges from land-based sources *directly* into the sea are concerned, the discharge must not cause a common law nuisance, even where the right to do so exists under a charter or by prescription, or immemorial usage. The main difficulty in seeking to prevent or restrain such a discharge lies in evidentiary proof that a person, or persons, are affected by it. At the same time, however, the Sea Fisheries Regulation Act 1966 enables the local District Sea Fishery Committee to make local by-laws to restrain discharges of 'solid or liquid substances'.

In turning to the present range of statutory provisions, it must be emphasised that, due to the brevity of this commentary, complete accuracy is simply not possible. The reader should, therefore, regard this merely as an indication of the history and broad position in the UK, and should refer elsewhere[4] if seeking a precise statement of law under each of the Acts referred to. For this reason, footnotes as to statutory section numbers etc. are deliberately omitted, since they should, in the author's view, be supported by further explanation if a degree of mis-interpretation of statutory wording is to be avoided.

As early as 1863, the Alkali, etc., Act (England and Wales) made it an offence to release chemical waste from a 'scheduled industry' (i.e. included in a schedule in regulations made under the Act, which could be extended by the responsible Minister), whether into water or other-wise, 'so as to cause nuisance', the offence being punishable by a fine. The current principal Act is now the Alkali, etc. Works Regulation Act 1906. A defence that the 'owner' of the industrial plant has exercised 'due diligence' exists; and the current maximum financial penalty is relatively light, although it can be levied on a continuous daily basis if the release of chemical waste does not cease. The 1863 Act also set up the Alkali Inspectorate (now the Alkali and Clean Air Inspectorate) to enforce the regulations. However, a private citizen may also initiate the preventative or primitive process by lodging a complaint with his local sanitary authority (i.e. District Council) who will then inform the Secretary of State for Health and Social Security, under whom the Inspectorate functions.

The Public Health Act 1936 defines a 'statutory nuisance' in a number of specific ways,[5] and applies it, *inter alia*, both to watercourses and to still water where the discharge results in the water being so foul as to 'prejudice health' or constitute a nuisance; or where a watercourse is so choked or silted up as to obstruct or impede the flow in such a way as to be either a nuisance or to give rise to conditions prejudicial to health. The sanctions applicable are, however, extremely limited, and are not appropriate to control of serious industrial pollution. This is not to say that very effective instruments do not exist elsewhere under the same Act, and under other more recent statutes. It is, in fact, a potent weapon that the Public Health Act 1936, in ss. 91-100, as amended by the Public Health (Recurring Nuisances) Act 1969, provides that a local authority may serve an *abatement notice*[6] on 'a person by whose act, default or sufference the nuisance arises or continues', or upon the building owner of nuisance arises from a structural defect. If, after the order, the nuisance further continues, the local magistrates' court may make a *nuisance order*[7] prohibiting a recurrence of the nuisance and imposing a small fine for each day's subsequent continuance.[8] The most significant factor is, however, that, after a nuisance order has been made, the local authority may enter and abate the nuisance, and subsequently recover the entire cost of abatement from the polluter.[9] In addition, the local authority may take proceedings in the High Court for abatement or prohibition of a statutory nuisance, in which event it may be significant that a person who elected to ignore, say, an order of prohibition could, at the discretion of the court, be imprisoned for contempt of court until such time as he was deemed to have purged that contempt.[10] The Act, in Part XI, also elsewhere provides specifically for small fines and powers to abate nuisances in connection with watercourses, ditches, ponds, etc., where they are 'so foul or in such a state as to be prejudicial to health or a nuisance'; or where a watercourse is so choked or silted up as to obstruct or impede the proper flow of water and thereby cause a nuisance, or give rise to conditions prejudicial to health,[11] but, again, the more significant provisions relate to the powers of the local authority to take positive steps to rectify matters, or to require the culverting of watercourses or ditches.[12]

However, although the foregoing controls still exist, a far more effective means of water management and water pollution control is now provided under the Water Act 1973 and Part II of the Control of Pollution Act 1974, together with (in certain respects) the Health and Safety at Work, etc., Act 1974.

The Water Act 1973 provided for the setting up of ten regional Water Authorities as part of a long-term national plan to develop systems of water conservation, purification and supply up to the year 2001 AD. These Water Authorities are autonomous bodies. Furthermore, they are run by non-elected boards, the members of which are appointed by the Secretary of State for the Environment and the Ministry of Agriculture, Fisheries and Food, jointly.

The responsibilities of each Water Authority specifically include control of drainage, construction of reservoirs and supply of water to consumers. The drainage control relates to virtually all types of drainage and includes ground water (from substrata), land drainage (agricultural field drainage), municipal sewage systems, industrial drainage and even sewage waste discharged from river boats of all sizes. However, a large part of this management responsibility in regard to public sewers and other localised aspects of drainage is contractually delegated to local authorities, such as District Councils or Metropolitan County or District Councils. The supply of water is delegated to statutory water companies.

The national plan for water includes many new and radical developments, such as the construction of bünded reservoirs in locations best suited to flexibility of conservation and supply (since bünded reservoirs avoid the need to find suitable valleys to flood, etc.); and it also visualises the increasing use of rivers and other waterways as 'longitudinal reservoirs' in which water can be both conserved and transported, the final purification being carried out at or near the point of destination.

A more thorough discussion can be found in the writer's earlier work,[4] but reference should be made here to certain more specific aspects of the Water Act. However, it can briefly be stated that each regional Water Authority has a statutory duty to provide for the conservation of water,[13] and to secure the proper use of water resources in their area, including the redistrubution of water within the area and the 'augmenting' of supplies—such as by use of desalination plant. The Water Authority is responsible for ensuring provision of a water supply[14] of sufficient volume and wholesomeness, subject to this being achievable 'at a reasonable cost'. A final decision as to whether or not such cost is reasonable in a given case can be made by the Secretary of State for the Environment. The duty to supply water can be met locally by delegation to a statutory water company.[15] These statutory water undertakers have power to supply water to places outside their 'limits of supply' through agreements made by the Water Authority for their region, but subject to the right of the Secretary of State (i.e. Environment and/or Agriculture, etc.) to intervene.[16] The provision of sewerage and sewage disposal services is now the responsibility of each Water Authority,[17] and this will include such facets as control of the design[18] of a sewer or drain within the general sewerage system operated by a local authority, and control of substances likely to injure sewers or drains, including chemical refuse or waste steam, petroleum spirit (including oil and any product of petroleum or mixture containing petroleum).[19] In fact, the responsibility – and thus the ultimate right of control – for, in effect, *all* aspects of sewerage and sewage disposal services lies with the regional Water Authorities.[20] Clearly, this does not mean that District Councils, County Councils, etc. must obtain specific approval for all their acts or decisions in regard to sewage, but rather that the Water Authority may step in and require positive steps to be taken, systems to be altered, or practices to cease. The statutory means of control of industrial pollution is, however, vested in the Water Authorities, not under the Water Act 1973, but by virtue of Part II of the Control of Pollution Act 1974, as will be seen later.

The Water Authority has also a responsibility to maintain, improve and develop the salmon fisheries, trout fisheries, freshwater fisheries and eel fisheries in the areas in which they exercise functions under the Salmon and Freshwater Fisheries Acts of 1923 and 1972. For this purpose, they must set up and maintain advisory committees with knowledge of the fishing in local rivers, estuaries, etc.[21] One should, perhaps, add that the Water Act requires Water Authorities to make considerable provision for recreational activities related to water,[22] and for nature conservation and amenity,[23] which can, of course, also have a direct bearing upon land-based marine pollution in river estuaries.

Turning to the Control of Pollution Act 1974, Part II: Pollution of Water, this Act provides wide powers of pollution control, which are exercised primarily by Water Authorities (although in certain respects by the Alkali and Clean Air Inspectorate), but all subject to ultimate control by the Secretary of State for the Environment.

The 1974 Act makes a deliberate distinction between: (a) (under s. 31) offences relating to the entry of any poisonous, noxious or polluting matter into any stream or *'controlled waters'* (*post*) or any 'specified underground water' (*post*)[24] (controlled waters and specified ground water are collectively referred to as 'relevant waters'); or entry of 'any matter' into a stream so as to tend to impede the proper flow of the stream[25] or of any 'solid waste matter to enter a stream or restricted waters'.[26] The essence of this is that liability will be deemed to exist, subject only to certain exclusions (as will be explained later); and (b) (under s. 32) the second category relates only to *'trade and sewage effluent'*, and the provisions as to the liability of a polluter are on a different basis, although situations may arise in which offences can fall under *both* categories.

Thus, with regard to all pollutants under s. 31 (*i.e. pollutants other than trade and sewage effluent*), it is an irrebuttable offence to cause or knowingly permit any poisonous, noxious or polluting matter to enter any stream or controlled waters or any specified underground water unless a disposal licence or a consent has been established.[27] In this context *'controlled water'*[28] means the sea within three nautical miles from any point on the coast measured from the low-water mark of ordinary spring tides, together with such other parts of the territorial sea adjacent to Great Britain as may be prescribed by the Secretary of State, and any other tidal waters in Great Britain, which latter will include enclosed docks or parts of the sea which may be specified by the Secretary of State by Order. The expression *'underground water'*[28] means any underground water specified in a document prepared by a Water Authority in a form prescribed for the purposes of this definition as water which is *used, or expected to be used*, for any purpose.

It is also an offence to cause or knowingly permit *'any matter'* to enter a *'stream'* so as to tend (either directly or in combination with other matter, which the offender or any *other* person causes or permits to enter the stream) to impede the proper flow of the stream in a manner likely to lead to a substantial aggravation of pollution due to other causes.[29]

A *'stream'*,[30] for the purposes of this Act, includes: any river, water-course or inland water, whether it be natural or artificial, or above or below ground, *except* a lake or pond which *does not discharge* into a stream (thus it presupposes that 'inland water' must link with other water!), nor into any sewer vested in the Water Authority (i.e. not a private sewer), nor into tidal waters.

The presumption of guilt under s. 31 will *not*, however, apply:

(i)   where the Water Authority has granted a 'disposal licence' (*post*);[31]
(ii) where consent has been given by the Secretary of State;[31]
(iii) where discharge has been earlier authorised under the Water Act *1945* (s. 34) or the Water (Scotland) Act *1946* (s. 50), or

under a local Act;[32]
(iv)   where a licence has been granted under the Dumping at Sea Act 1974;[33]
 (v)   where the entry of the polluting or other matter is in accordance with 'good agricultural practice', *except* where a notice has been served to the effect that the entry must not occur;[34]
(vi)   where the entry results from an emergency and in order to avoid danger to the public (the entry being notified to the Water Authority as soon as reasonably practicable after the event);[35]
(vii)   in cases where the matter will fall to be dealt with under s. 32 and not s. 31.[36]

The provisions of s. 31 do also extend to certain other aspects of liability, and there is, further, the question of case law in regard to the interpretation of 'to cause' the entry of polluting matter, as defined by the House of Lords in *Alphacell Limited* v. *Woodward* [1972] 2 All E.R. 475, and in which it was held that, if causation could be proved, it was then unnecessary for the prosecution to prove knowledge, intent or negligence on the part of the defendants.

In addition to the specific restrictions imposed by s. 31, the same section also empowers Water Authorities and (separately) the Secretary of State to make regulations and also by-laws for specified areas of activities.[37] The penalties applicable will probably be as stated in those regulations or by-laws; but, in the absence of such stated penalties, the general provisions of s. 31 as to penalties will apply.[38] In regard to such by-laws, these are relatively light, and do not extend to any power of imprisonment.

As already stated, offences arising out of *discharges of trade and sewage effluent* and other 'matter' under certain specified circumstances are covered by s. 32. It is an offence to cause or knowingly permit discharge of any trade effluent or sewage:

 (i)   into any stream or controlled waters or specified underground water (which are collectively referred to in the Act as 'relevant waters');[39] or
 (ii)   from land in Great Britain through a pipe into the sea outside controlled waters (for example beyond the 3-mile limit (*ante*));[40] or
(iii)   from a building or plant on to or into any land or into any lake, loch or pond which does not discharge into a stream[41] (i.e. this

is a stricter provision in regard to ponds, etc., than under s. 31).

In addition, s. 32 provides that it is an offence to discharge *any matter, other than trade or sewage effluent, from a sewer*[42] (as defined by the Public Health Acts) or from a drain[43] (as so defined). To put this as simply as possible, the relevant sewers or drains will be those connected with buildings or from yards connected with buildings. And, lastly, it is an offence to discharge *any matter, other than trade or sewage effluent*, into 'relevant waters' (*ante*) *from* 'a drain which a highway authority or other person is entitled to keep open'.[43]

In the case of offences under s. 32, penalties can be imposed on summary conviction of a fine and/or imprisonment not exceeding three months; or, on conviction on indictment, of a fine and/or imprisonment not exceeding two years.[44]

With regard to *consents for discharges* (other than from ships and other vessels[45]) a Water Authority and/or the Secretary of State may grant these in regard to a *specific* discharge.[46] Applications for such consents must provide details as to the precise point of discharge, and the maximum quantity and rate of discharge.[47] Provisions are made for appeal to the Secretary of State on a basis of deemed refusal after three months delay by a Water Authority.[48] Before a consent is granted, a notice of the application must be published in two successive weeks in a newspaper circulating in the area and in any other areas in which there are any streams likely to be affected.[49] In addition, such notice must be published in the *London Gazette*,[49] and copies of the application must be sent to each local authority in whose area a discharge is to be made.[50] Where the discharge relates to controlled waters, copies must also be sent to the Minister of Agriculture, Fisheries and Food.[50] Written representations (including objections) may be made within six weeks to the Water Authority,[51] and, where the Water Authority proposes to grant consent, they must serve notice of the proposal on persons who have made representations, indicating that they may, within 21 days, request the Secretary of State to make a direction in the matter.[52] In so doing, the Secretary of State must, with certain exceptions, first hold a *local inquiry*, which may or may not be in public.[53]

Consents must specify a *period* during which a notice of revocation or alteration etc. cannot normally be served, and this period shall be a 'reasonable period of not less than two years', although 'emergency' situations may justify earlier revocation or alteration.[54] Where consent is granted subject to *conditions*, the Water Authority may control the place, design and construction of the outlets; the nature, composition, temperature and rate of discharges and the periods during which they may be made; the provision of facilities for taking samples (including use of manholes, inspection chambers, observation walls and boreholes); the provision, maintenance and testing of meters and other apparatus; the keeping of records; the making of returns and provision of other information to the Water Authority about the nature, composition, temperature, volume and rate of discharges; and also any steps to be taken for preventing the discharges from coming into contact with any specified underground water.[55] (*Note*: the foregoing explanation is approximate only, and is essentially intended by way of general description. If the reader wishes to take any practical steps related to the Control of Pollution Act 1974, and other related legislation, he should refer at least to the work indicated in footnote 4.)

It should be stressed that separate provisions are made for *control of discharges of trade effluent into public sewers* and that these make practical allowance for the earlier statutory authorisation of discharges of trade effluent under s. 4 of the Public Health (Drainage of Trade Premises) Act 1937. These authorisations were granted by local authorities (i.e. prior to reorganisation of local government in England, the councils of county boroughs or non-county boroughs, or urban or rural districts).[56] In broad terms, it can be said that such discharges will have been limited to a specified quantity and rate of discharge, which cannot now be increased without further consent from the Water Authority. Naturally, if at any time the discharge is found to contain *toxic* chemicals which are subject to any other statutory prohibition, the consent can be withdrawn or (possibly) suspended, or an express consent under the 1974 Act (*ante*) substituted.[57] (*Note*: again, this description is in outline only; see *Note* above.)

Thus, as to land-based marine pollution in the UK, it will be seen that the statutory provisions do not seek to offer a separate code, since it is realised that by far the greater part of pollutants entering marine waters emanate from rivers and their estuaries.

Finally, however, reference should be made to the specific legislation in regard to oil pollution from drilling rigs and other installations and pipelines, since these are deemed to represent land-based sources in terms of the definition applied under the European Community Directive on land-based marine pollution (see Chapter 10). In UK legislation, escapes and leakages are covered (in England) by the Prevention of Oil Pollution Act 1971, and provision is made for an unlimited fine on conviction or indictment. Specific defences, however, apply in that the polluter may establish that, where the escape or leakage was due to damage *not caused by* the polluter, his servants or agents, and reasonable steps to prevent, stop or reduce the escape were taken as soon as practicable; or alternatively that, where delay did occur, it was not due to lack of 'reasonable care'.

3 COUNTRIES HAVING AN ADVANCED INDUSTRIAL ECONOMY BUT CONSIDERABLE LAND SPACE: E.G. THE UNITED STATES OF AMERICA

As already indicated, the USA will be taken as an example both of an advanced economy reliant upon industrial production but having a low population density relative to the total land mass; a federal system of government as distinct from the system of international co-operation adopted in the European Community.

It is, perhaps, not unfair to say that town and country planning in the United States has not yet reached the levels of sophistication achieved in Western Europe. This is no doubt due in part to the greater spirit of individual freedom and enterprise (more readily supported by the vast natural resources still untapped in the Americas) and in part to the very large areas of land under the control of the governments of individual States and of the Federal Government. The effect has, perhaps, also resulted from the systems of government within many of the individual States, which require that the elected State Legislature meet only infrequently (in some cases for two months every two years); and that day-to-day State government be conducted by the State Governor assisted by Commissioners responsible for specific functions. Thus, the pressures exerted by the electorate may, so far as lesser items of planning control are concerned, tend to be reduced in their immediate effect. In addition, the responsibilities of county authorities within each State may tend to lack clear definition. As a result, there are instances of acute water pollution in rivers in the US—even to the extent of rivers recurrently catching fire—which should have been avoidable, given the land availability in the greater part of the federal territory.

As a result of inadequate control of industrial pollution, and of certain other developments having very marked effects upon local environments, individual citizens or associations have in recent years pursued common law and statutory rights in the courts, seeking injunctions to restrain either pollution or intended industrial or other development.

It is not possible to discuss here the wider problems of planning and land use, but it should be observed that the Annual Report to the House of Delegates of the American Bar Association for 1974 included recommendations from the Association's Special Committee on Environmental Law (Report No. 106) which stated that:

> At present, decisions for industrial site selection are frequently made on an *ad hoc* basis without the benefit of integrated natural resource planning by state and local units of government. Since nearly all major industrial facilities have an impact on state and regional interests as well as local interests, there is an obvious need for local planning to be integrated within the framework of statewide planning. The planning process should include use of water resources as well as land, and should determine the general compatibility of various kinds of industrial development within different regions of the State. If natural resource planning on the state level is implemented in an effective manner, then many of the problems now encountered in industrial site selection will be resolved at the planning stage. Opportunity for public participation at all stages, levels and phases of planning is essential.
>
> Legislation pending in Congress [1974] might prompt land use planning programs on the state level. But the need for natural resource planning on the state level is too urgent to await federal initiatives, and the States should implement planning programs on their own initiative.

The Committee recommendation and the resolution adopted by the American Bar Association in regard to this situation included:

> Each State should provide for comprehensive and co-ordinated statewide planning to assure wise and prudent use and conservation of natural resources, and should provide planning criteria for evaluating proposed uses of those resources in relation to developmental objectives and environmental values, including bio-physical, social, cultural and economic,

and it further proposed that this be achieved by each State setting up an Industrial Siting Council (ISC) having jurisdiction over such matters.

However, it should be noted that, at the time of this initiative by the American Bar Association, there already existed the National Environmental Policy Act of 1969; and (in the particular context of the present chapter) the Coastal Zone Management Act of 1972 and the Federal Water Pollution Act of 1972 (since amended, *post*).

For present purposes, it may suffice to say that the National Environmental Policy Act (NEPA) came into force on 1 January 1970 and had three major aims:

> (1) to declare protection of environmental quality to be a national policy and provide a mandate to all Federal agencies to effect that policy; (2) to create a Council on Environmental Quality to insure that the mandate is carried out; and (3) to establish a set of 'action forcing' procedures requiring an environmental impact statement [EIS] on any proposed major Federal action which could significantly affect the quality of the environment.[58]

It should be observed, however, that the precise requirement that there be an EIS relates only to *federal agencies* and that the wider responsibility for environmental management, as applicable to all levels of federal, State and local government, is referred to merely as a principle, without definition of detailed legal powers or duties.

Turning to matters of land-based marine pollution as such, it is convenient to deal first briefly with the Coastal Zone Management Act of 1972. This formally recognises (section 302) the importance to the national interest of 'effective management, beneficial use, protection and development of the coastal zone'. For the purposes of this Act, the *coastal zone* means:

> coastal waters (including the lands therein and thereunder), strongly influenced by each other and in proximity to the shorelines of the several coastal states, and includes transitional and inter-tidal areas, salt marshes, wetlands and beaches. The zone extends, in Great Lakes waters, to the international boundary between the US and Canada and, in other areas, seaward to the outer limit of the US territorial sea. The zone extends inland from the shorelines only to the extent necessary to control shorelands, the uses of which have a direct and significant impact on the coastal waters. Excluded from the coastal zone are lands the use of which is by law subject solely to the discretion of or which is held in trust by the Federal Government, its officers or agents.

It will thus be seen that the 'coastal zone' does not equate precisely to the European Community 'coastal area' as at present provisionally being considered by the European Commission, which contemplates a much more extensive planning control over the coastal hinterland (see Chapter 10). The US Act of 1972 has in mind (s. 302) the demands, *inter alia*, of population growth, industry, recreation, extraction of

minerals and fossil fuels, navigation, waste disposal, fisheries and harvesting of shellfish and other marine resources, which demands are having an
adverse effect. And, in consequence, the Act declares that it is national
policy 'to preserve, protect, develop and, where possible, to restore or
enhance the resources of' the coastal zone; to encourage and assist
States and local governments in this respect; and to require federal
agencies to co-operate and participate with State and local governments
and regional agencies in effecting the intent of the Act. However, this
again represents a statement of general policy and does not provide
precise and mandatory legislation.

However, the requirements of federal legislation as to pollution of
both fresh water and sea-water are now laid down in the Federal Water
Polluting Control Act of 1972,[59] as amended by the Clean Water Act of
1977.[60] The objective of the Act is to 'restore and maintain the chemical,
physical and biological integrity of the Nation's Waters', by eliminating
the discharge of pollutants into such waters by 1985.[61] (*Note*: the Act
refers to *'navigable waters'*, but this is defined as 'the waters of the
United States, including territorial seas',[62] and would thus appear to
include the headwaters of rivers which may not, in fact, be navigable.)
It also provides for an interim goal of water quality for the protection
and propagation of fish, shellfish, wildlife and recreation by 1 July
1983; and for the immediate (1972) prohibition of discharge of toxic
pollutants 'in toxic amounts'. To this end, federal financial assistance is
provided for construction of *publicly* owned waste treatment works, and
specifies that it shall be policy to develop 'areawide waste treatment
management planning processes' so as to assume adequate control of
sources of pollutants in each State. The Act also enables a major research
and demonstration effort in the development of technology for pollution
control or elimination.[63]

At the same time, however, the Act recognises and relies upon the
primary rights and responsibilities of individual State governments in
achieving the aims and requirements laid down. A general reading of
the Act makes it clear that individual States must implement specified
procedures and practical steps within definite time limits, and that the
Federal Government has power to act in default; but, equally, that
reasonable elements of discretion remain (as in the case of most Federal
legislation) with State Governors. These State rights are supportable by
action in the courts on the initiative of a State government.

The Federal Water Pollution Control Act (as amended) is of considerable length and complexity, and it is possible to give here only an
indication of the more significant features.

The Act provides[64] that it shall be administered by the *Administrator of the Environmental Protection Agency* (subsequently referred to as 'the Administrator') (of the Federal Government). It requires the Administrator to investigate for and draw up comprehensive programmes in connection with other federal agencies, State water pollution control agencies, inter-State agencies, municipalities and industries involved in regard to all 'navigable waters', i.e. fresh water, sea-water and *ground waters*, etc.[65] It also provides (with certain reservations) for the Administrator to approve federal agency projects for water conservation, storage and control, including hydroelectric power projects.[66]

State governments have been required to set up State water pollution control (or planning) agencies and to prepare a comprehensive pollution control plan for each 'basin'. A *basin* (as defined) 'includes, but is not limited to, rivers and their tributaries, streams, coastal waters, sounds, estuaries, bays, lakes, and portions thereof, as well as the lands drained thereby'.[67] The *plan* for each basin is to be consistent as to water quality standards, effluent and other limitations, and thermal discharge regulations made or approved by the Administrator;[68] is to recommend treatment works, means of collection, storage and elimination of pollutants in regard both to municipal and industrial use of such works;[69] and is to define the means of maintaining water quality (including financial proposals).[70] Where necessary, States are to co-operate or to seek to develop uniform laws as to water pollution control.

The Administrator is also required to establish national programmes for the prevention, reduction and elimination of pollution,[71] in co-operation with State and other bodies;[72] to initiate and promote research as to the most effective practicable tools and techniques of measuring the social and economic costs[73] and benefits of activities covered by the Act. He is also to develop effective and practical processes, methods and prototype devices for the prevention, reduction and elimination of pollution.[74] The research and studies are to relate to treatment of both municipal sewage and other water-borne wastes, identification and measurement of the effects of pollutants, 'including those pollutants created by new technological developments' and in regard to effects on water quality of 'augmented streamflows'.[75] These duties of the Administrator may be carried out by establishing field laboratories or research facilities, either directly or by making grants to other public or private agencies (including also colleges and universities);[76] and grants may be made for this purpose.[77] It should, perhaps, be observed that special reference is made to estuaries and to 'estuarine zones' (i.e. an environmental system consisting of an estuary and those transitional areas which are influenced or affected by water from an estuary).[78]

Grants are also authorised[79] for preparation of the control programmes of States and of inter-State agencies, where the programme and its cost are approved by the Administrator; and for scholarships and for the training, by institutions of higher education, of students in the design, operation and maintenance of treatment works and other facilities whose purpose is water quality control.[80]

Specific provision is made for federal grants to public bodies for the construction of treatment works and for systems of recycling of potential sewage pollutants through the production of agricultural, silvicultural and aquacultural products, and by other means; for reclamation of waste water; and for the ultimate disposal of sludge in a manner which will not result in environmental hazards.[81] Considerable detail is given as to the conditions applicable to these grants, but the general theme (as is the case throughout the Act) is that initiatives should be encouraged, provided that they are taken on a realistic basis and at realistic cost[81] – plans and specifications being previously submitted to the Administrator for approval.[82]

The Act also refers to *'areawide waste treatment and management plans'*,[83] the guidelines for which were required to be published as regulations by the Administrator, after consultation with the appropriate federal, State and local authorities, within 90 days of the coming into force of the 1972 Act. Such management plans are to contain alternatives for waste treatment management and be applicable to all wastes generated within the area involved. They are certified by the State Governor and submitted not later than two years after the planning process is in operation; or, if the agency was designated after 1975, the plan involved is to be submitted not later than three years after receipt of the initial grant (awarded from federal funds).[84] Such plans are to be for a twenty-year period and annually updated. They are to cover at least[85] the identification of treatment works necessary to meet the anticipated municipal and industrial needs; water-based recreation; the agencies necessary to operate the plan; costs in terms of economic, social and environmental impact; identification of *nonpoint sources of pollution*, including irrigated agriculture, run-off from marine disposal areas and from land used for livestock and crop production; mine-related sources of pollution (surface and underground); forms of 'construction activity related' sources of (water) pollution; salt water intrusion into rivers, lakes and estuaries as a result of reduction in fresh water flow from any cause, including irrigation, obstruction, ground water extraction or diversion; the disposal of all residual waste generated in the area which could affect water quality; and a process to control the disposal of pollutants on land or subsurface excavations within such area, so as to protect ground and surface water quality. It is, however, important to note that these provisions do *not* relate to *disposal of sewage sludge* where the sludge results from the operation of a treatment works[86] (including the removal of in-place sewer sludge from one location and its deposit in another location)[87] if this would result in any pollutant entering the navigable waters. Such disposal is prohibited[88] except in accordance with a permit issued by the Administrator (under s. 402). However, the Administrator is to make regulations, which are to include guidelines as to uses for, and disposal of, sludge etc., and a State government may apply for approval to administer its own scheme in accordance with s. 402 of the Act (*post*). Provision is also made for individual States to propose state-wide water quality standards and, if these are approved by the Administrator, to apply them throughout the State.[88]

The operation of an 'areawide waste treatment management planning process' (i.e. not the actual cost of treatment) was subject to a 100 per cent grant from federal funds, up to 1 October 1977 and, subsequently, to a 75 per cent grant.

It is significant, as a traditional method adopted in the United States where major civil engineering works are concerned, that the Secretary of the Army, acting through the Chief of Engineers, can provide advice or (in some cases) practical aid. Under the present Act he may, in co-operation with the Administrator, at the request of a State Governor, consult with and provide technical assistance[89] to any agency designated by the State to operate area-wide waste treatment; and substantial federal funds are allocated towards the cost. Equally, the Secretary of the Interior, acting through the Director of the United States Fish and Wildlife Service, provides technical assistance free of charge.[90] And the Secretary of Agriculture (with the concurrence of the Administrator), acting through the Soil Conservation Service and other agencies of the Department of Agriculture, may enter into contracts relating to the introduction of the best agricultural, etc., *management practices* in terms of reduction of water pollution.[91]

It should also be observed in regard to the planning, operation and maintenance of treatment works that the Administrator is required to carry out an annual survey,[92] and, second, (of importance) that there is a strict limitation upon the availability of grants for *sewage collection systems*, these being available only where replacement or major rehabilitation is required; or where a new system is essential in order to meet the water standards necessary under the Act.[93]

Turning to the important question of enforcement of water pollution control, the Act provides that *'effluent limitations'* shall apply to *point sources other than publicly owned treatment works*, which require the application of the *'best practicable technology currently available'*, as defined by the Administrator;[94] or that, where the point source discharge is into publicly owned treatment works, the discharge shall meet other standards specified by the Administrator,[95] which shall relate to pre-treatment and secondary treatment.

However, the Act then goes on to provide for certain categories of pollutant from point sources (other than public owned treatment works) where the standard is the *best available technology 'economically achievable'* . . . 'which will result in reasonable further progress toward the national goal of eliminating the discharge of all pollutants'.[96]

The standard required as 'economically achievable' is, however, specified in regulations issued by the Administrator, and the relaxation applicable in economic terms is confined to certain listed pollutants which fall under three general headings (given in s. 301(b)(2)(C),(D) and (F)). These are (i) all toxic pollutants listed in a *Table* prepared by the Committee of Public Works and Transportation of the House of Representatives[97] (*ante*); (ii) other pollutants given in a *list published* (and periodically reviewed) by the Administrator taking into account the *degree of toxicity, persistence, degradability and the usual and potential presence of organisms in any water, together with the importance of those organisms*; and (iii) other pollutants *not* included in or under defined categories and classes of point sources identified by regulations made by the Administrator as those to which *'the best conventional pollution control technology'* shall apply (as required by s. 304(a)(4)). Nevertheless, as a final saving clause, the Administrator is given discretion to modify the requirements[98] where a permit application has been made for a point source and the applicant can establish that the modification will (i) represent the maximum technology within the 'economic capability of the owner or operator' *and* (ii) will result in 'reasonable further progress towards elimination of the discharge of pollutants'. However, the Administrator is also required to seek the concurrence of the State government and to be satisfied that such a modification will not increase the burden of other point or non-point sources, and that it will not affect public water supplies and the protection of shellfish, fish, wildlife and recreational activities, or be 'reasonably anticipated to pose an unacceptable risk to human health or the environment because of bioaccumulation, persistency in the environment, acute toxicity, chronic toxicity (including carcinogenicity, mutagenicity or teratogenicity) or synergiatic propensities'.[99]

In the face of this web of reservations, a polluter may be expected to take the view that the question of economic cost will, in most cases, not be legally relevant.

However, so far as existing *discharges from publicly owned treatment works into marine waters* are concerned, the Administrator may (with the concurrence of the State government) issue a permit[100] if the applicant demonstrates at a public hearing that there is an applicable water quality standard (identifiable under s. 304(a)(6)) published by the Administrator and that the discharge, if the requirements are modified as requested, will not interfere with public water supplies or with shellfish, fish and wildlife, or with recreational activities, or increase the demands on any other point or non-point source. It is also a condition that all pre-treatment requirements will be met and that there will be no new or substantially increased discharge above the volume specified in the permit, if granted.[100] For these purposes, 'marine waters' include deep waters of the territorial sea or the waters of the contiguous zone[101] or 'saline estuarine waters where there is a strong tidal movement'.[100]

To achieve the aims already indicated, the Administrator will apply '*effluent limitations* (including alternative effluent control strategies) to any point source, or group of point sources'.[102] In order to do so, he must first issue a notice of intent and, within ninety days, hold a public hearing to consider the question of 'economic and social dislocation in the affected community', and equally, the economic and social benefits. If it can be demonstrated that the latter are not justifiable as against the dislocation, then the intended effluent limitation shall not be applied, or it shall be modified.[103]

In order to achieve the requirements of the Act, each State must adopt *water quality standards*[104] for all waters 'within the State' (i.e. *including* coastal waters) and submit these to the Administrator for approval.[105] If the Administrator finds these standards insufficient he will specify amendments and, if they are not adopted, he may then lay down the standards by regulations, which must be followed by the State.[106] In addition, whether the standards are as initially prepared by the State or are as modified by the Administrator, the State Governor, or the State water pollution control agency, must, at least once in every three years, hold public hearings for the purpose of reviewing these standards—any resultant revision being, again, subject to the approval of the Administrator.[107]

Each State is also required to identify those waters within its territories for which the effluent limitations 'are not stringent enough to implement any water quality standard applicable to those waters', and to apply a priority ranking to them for future action.[108] The same will apply to waters in which thermal discharges endanger shellfish, fish and wild-life.[109] In both cases (i.e. where pollution and thermal discharges exceed the limitations), the State shall estimate a maximum daily load of pollutant or thermal discharge so as to assure protection of shellfish, fish and wildlife, but may allow for seasonal variations, etc.[109]

In terms of forward planning, each State was initially required to submit to the Administrator within 120 days of the 1972 legislation, a *'proposed continuing forward planning process'*, which was consistent with the Act, to cover all 'navigable waters' within the State. These are to include (but are not limited to) effluent limitations and schedules of compliance (to be at least as stringent as the Act requires); area-wide waste management plans and basin plans; total maximum daily loads[110] for pollutants; procedures for revision; adequate authority for inter-governmental co-operation (i.e. State-federal and inter-State); adequate implementation, including schedules of compliance, for revisal of any new water quality standards; controls over disposal of all residual waste from water treatment processing; and an inventory and ranking, in order of priority, of need for the construction of waste treatment works.[111]

The expression *'treatment works'* is defined as:[112]

any devices and systems used in storage, treatment, recycling and reclamation of municipal sewage or industrial wastes of a liquid nature to implement section 201 of this Act,[113] or necessary to recycle or re-use water at the most economical cost over the estimated life of the works, including intercepting sewers, outfall sewers, sewage collection systems,[114] pumping, power and other equipment, and their appurtenances; extensions, improvements, remodelling, additions, and alterations thereof; elements essential to provide a reliable recycled supply such as standby treatment units or clear well facilities; and any works, including site acquisition of the land that will be an integral part of the treatment process (including land use for the storage of treated wastewater in land treatment systems prior to land application), or is used for ultimate disposal of residues resulting from such treatment.[115]

In addition, however, to the above, *'treatment works' also means:*

'*any other method* or system for preventing, abating, reducing, storing, treating, separating, or disposing of municipal waste, including storm water runoff, or industrial waste, including waste in combined[116] storm water and sanitary sewer systems'.[116]

It is not surprising, in view of the complexity and detailed requirements of this Act, that provision is made for the Administrator to obtain and provide information, and to lay down *guidelines and criteria.*[117] *The criteria for water quality* are to reflect accurately the latest scientific knowledge

(A) on the kind and extent of all identifiable effects on health and welfare including, but not limited to, plankton, fish, shellfish, wildlife, plant life, shorelines, beaches, esthetics [!], and recreation which may be expected from the presence of pollutants in any body of water, including ground water; (B) on the concentration and disposal of pollutants or their byproducts, through biological, physical and chemical processes; and (C) on the effects of pollutants on biological community, diversity, productivity, and stability, including information on the factors affecting rates of eutrophication and rates of organic and inorganic sedimentation for varying types of receiving waters.[118]

As to *non-point sources of pollution*, the *guidelines*[119] relate (i) to identifying and evaluating their nature and extent; and (ii) to processes, procedures and methods to control pollution resulting from –(a) agricultural and silvicultural activities, including run-off from fields and crop and forest lands; (b) mining activities, including run-off and siltation from new, currently operating and abandoned surface and underground mines; (c) all construction activity, including run-off from facilities resulting from such construction; (d) the disposal of pollutants in wells or in subsurface excavations; (e) salt-water intrusion resulting from reductions of fresh water flow from any case, including obstruction and diversion; and (f) changes in the movement, flow or circulation of any navigable waters or ground waters, including changes caused by the construction of dams, levees, channels, causeways or flow diversion facilities. All this information is to be published in the Federal Register and otherwise made available to the public.

Further guidelines relate to suitable pretreatment of the different categories of pollutants (revised annually), test procedures and procedures for monitoring, reporting, enforcing and funding all necessary steps required under the Act.[120]

In addition to these general guidelines, '*national standards of performance*' are laid down for categories of sources listed by the Administrator (under s. 306(b)(1)(A)); and State laws are to ensure that these are met, and that no *new source* shall operate in violation of them. These categories of sources (which are listed, as the minimal requirement under the Act, in footnote 121) include a wide range of types of process, from chemical manufacture to pulp and paper mills and feedlots. However, where any interested person contests the validity of the Administrator's decision to include a category in the list, he may apply to the Circuit Court of Appeals of the US for his Federal District.[122]

As already mentioned, the Administrator is also required (under s. 307) to lay down *toxic and pretreatment effluent standards*, which require that the 'best available technology economically achievable' be applied.[123] Proposed effluent standards (or also prohibitions) are to be published in the Federal Register and will not come into force for sixty days after publication, during which time written comment will be received from interested parties (persons). If such a person so requests within *thirty days*, the Administrator must hold a public hearing at which oral and written representations may be made, and 'such cross-examination as the Administrator determines is appropriate on disputed issues of material fact' may take place. A verbatim record is to be kept.[123] After consideration of such evidence, the Administrator will promulgate the relevant effluent limitations and standards (or prohibitions) — revised if he so decides — but his decision may be subject to judicial review, on the basis that it is contrary to the balance of the evidence adduced at the public hearing.

To ensure the practical application of these requirements, the Administrator has a right of entry on to relevant premises,[124] and a right of inspection of records.[124] He may also require an owner, or operator of any *point source* to use and maintain monitoring equipment, take samples and keep records.[125] Any information so obtained shall be available to the public *except* that 'any person' ('person' includes a corporate body, etc., and will include, for example, a manufacturer) may prove that, if made public it would divulge methods or processes entitled to protection as *trade secrets.*[126] It is of interest that the initiative to seek confidentiality lies with the manufacturer and that the Act does not lay down any initial period in which such initiative is to be taken and during which information will not be revealed. The usual provision is made for each State, as an alternative, to develop and submit State procedures to the Administrator for approval;[127] but the Act does not make specific reference to protection of trade secrets under such procedures, although, no doubt, the Administrator would require this provision before approving the State system.

Provision is made in the Act (s. 309) for *federal enforcement.* If the Administrator finds that any person is in violation of the conditions or limitations applicable to a permit issued under the Act, he will notify the State, and if that State fails to take appropriate enforcement action within 30 days, he shall issue an order requiring such person to comply *or* shall bring a *civil action* in the District Court (a *Federal* Court) against that person for 'appropriate relief, including a permanent or temporary injunction'.[128] In cases where the person *wilfully or negligently* violates the relevant sections of the Act or any permit condition or limitation, he may be punished by a fine, or by imprisonment for not more than one year, or both.[129] Where the conviction is for a violation committed after a first conviction, the penalties are doubled. In contrast, where the person has acted in good faith and has made a contractual or other commitment, an extension to the period for compliance may be granted.[130] Where a *municipality* is a party to a civil action brought by the Federal Government, the State of which it is part shall be joined as a party, and the *State* shall be liable for payment of any judgment or any expenses incurred as a result of complying with the judgment to the extent (if relevant) that the laws of the said State *prevent* the municipality from raising revenues for this purpose.[131]

In cases where the violations of permit conditions or limitations 'are so widespread that such violations appear to result from failure of the State to enforce' them, the Administrator must inform the State; and, if such failure continues after 30 days, he must give public notice of his finding. If the State still fails to enforce the requirements, the Administrator may issue an order that they comply, *or* he may bring a civil action.[132]

The foregoing provisions of the Act have been discussed at some length on the principle that by far the greater part of marine pollution emanates from rivers and other land-based sources, and the approach adopted in the entire Act appears to allow for such an assumption. Thus, in providing for *oil and hazardous substance liability* (s. 311), the Act covers control of oil pollution by means of provisions almost the entirety of which apply both to fresh water and sea-water; and it is stated that: 'it is the policy of the United States that there should be no discharges of oil or hazardous substances into or upon the navigable waters of the United States, adjoining shorelines, or into or upon the waters of the contiguous zone . . ,'[133] or in connection with the outer continental shelf, etc. (It should be recalled that 'navigable waters' is defined in such a way as to cover all waters of the US.)

The Act refers to the liability of owners or operators of vessels, on-shore facilities and off-shore facilities; and it is significant that 'on-shore facilities' are defined as 'any facility (**including,** but not limited to, motor vehicles and rolling stock) of any kind located in, on, or under, any land within the United States other than submerged land'.[134]

The Act prohibits discharges of oil or hazardous substances (discharges being defined as including, but not limited to, any spilling, leaking, pumping, pouring, emitting, emptying or dumping)[135] in 'harmful quantities'.[136] The quantities deemed to be harmful are specified by regulation, and are assessed in regard to locations and conditions and to public health or the health or welfare of 'fish, shellfish, wildlife, and public and private property, shorelines and beaches'.[137]

It is not appropriate to this chapter to discuss control of discharges from *vessels at sea*, but it should be noted that these will be, equally, covered by the general provisions of s. 311 of the Act, including certain more specific subsections and paragraphs; and that 'marine disasters' are covered, *inter alia*, by s. 311(d).

Any owner, operator or person in charge of any on-shore or off-shore facility (but not a vessel) from which oil or a hazardous substance is discharged should be assessed for a civil penalty, following a hearing by the Secretary of the department in which the Coast Guard is operating; and this penalty will relate to the size of the polluter's business, his ability to continue that business, and the gravity of the violation.[138]

There is also specific authorisation enabling the President to act to remove such oil or substances, either when it is discharged or 'if there is a *substantial threat of discharge*'.[139] This latter provision is of particular significance and value, but it may present problems in terms of establishing the validity of the assessment of a 'threat of discharge'.

The Act also provided for the immediate preparation and publication of a *National Contingency Plan* for removal of oil and hazardous substances in pursuance of the aforesaid power given to the President. This Plan covers co-ordinated action to mimimise damage from oil and hazardous substance discharges, and this includes their containment, dispersal and removal. It has set up and equipped an organisation, including a strike force, and a system of surveillance under a national centre. Due provision is made for co-operation with, and delegation to, State governments[140] in these functions.

In regard to both *on-shore facilities and off-shore facilities*, where discharges occur an automatic liability to reimburse the US Government up to $50 million will apply, subject only to specified defences, namely: act of God, act of war, negligence on the part of the US Government; or an 'act or omission on the part of a third party without regard to whether any such act or omission was or was not negligent'; or any combination of the foregoing factors.[141] Perhaps it is reasonable to observe that circumstances could arise in which there is a chain of causation originating in a *non*-negligent act on the part of the US Government—for which the Act would appear to make no provision. In terms of liability where the discharge is solely due to an act or omission on the part of a *State government*, no doubt the defence of 'act or omission on the part of a third party' can apply; but this will be subject to any State laws to the contrary, since no specific provision is made in the Clean Water Act.

The liability for reimbursement of the Federal Government will extend not only to cost of removal, etc., of the pollutant oil or hazardous substance, but will also cover the cost of *replacing or restoring natural resources*,[142] which could extend to, for example, the restocking (if possible) of shellfish beds; and, presumably, recolonising areas with wild birds or breeds destroyed by oil pollution. In terms of this total cost, a specific federal right as against any third party is reserved by the Act.[143]

The *liability of the Federal Government* for its own 'facilities' (e.g. Government land, buildings, installations, etc.) is specifically referred to in the Act. S. 313 requires that the same provisions in regard to water pollution as apply to any 'person' shall equally apply to each officer, agent or employee of the Federal Government when in performance of official duties.[144] However, the officer, agent or employee will not be personally liable for 'any civil penalty arising out of the performance of his official duties for which he is not otherwise liable'.[144]

Subject to the overall requirement of meeting pollution control standards, limitations, prohibitions and other requirements of the Act, individual effluent sources from federal facilities may be exempted by the President, *except for* those covered by ss. 306 or 307 of the Act (*ante*), i.e. those categories of sources listed in regulations made under s. 306 (which range from chemical manufacturing to pulp and paper mills, and feedlots); and those sources emitting toxic pollutants listed by virtue of s. 307. In other words, the President's discretion is excluded from sources producing the more dangerous or inconvenient types of pollutants, other than sewage. And, in any event, the decision is supportable only if it is deemed to be 'in the paramount interest of the United States', and no exemptions can be granted due to lack of budgetary appropriation, unless the President has specifically requested such appropriation and the Congress has failed to make it available. However, exemption will, in any event, be allowable for essentially military

reasons (*ante*). On rather similar lines, the Administrator is given power
to permit discharges of specific pollutants under controlled conditions
if this is associated with an approved *aquaculture project* to be carried
out under federal or State supervision.[145]

The general machinery for operation of the controls under the Act is
that licences and permits are issued through an authorised agency in
each State, subject to certification by the State that any standards
required by federal regulations, etc., are met; and any such certification
and issue of a licence or permit must be notified to the Administrator.
The requirements in this respect are fairly complicated and are laid
down in s. 401 of the Act. They are also linked with the National
Pollutant Discharge Elimination Scheme provided for in s. 402, which
relates them to the specific provisions of the earlier sections of the Act
already discussed; but provision is made[146] for the Administrator to
issue a permit for the discharge of any pollutant, or combination of
pollutants, after holding a public hearing, *provided that* the discharge
will actually meet the requirements of ss. 301, 302, 306, 307, 308
and 403 of the Act. These are considered earlier in this chapter, save
for s. 403, which relates to ocean discharge criteria, and is not deemed
relevant to land-based pollution.

Special permits are also available for dredged or fill material for
disposal at specified sites, these being issued by the Secretary of the
Army, acting through the Chief of Engineers; but the Administrator
has power to prohibit specification of any given site. Where an individual
State desires to administer its own permit programme for discharge of
dredged or fill material, the programme must be first submitted for
approval by the Administrator.[147]

As further aids to the effective application of the Act, the Federal
Government may, through the Administrator, award certificates or
plaques to industrial organisations (presumably including manufacturers)
and to political subdivisions (e.g. counties and municipalities).[148] In
addition, a Water Pollution Control Advisory Board[149] has been set up
under the chairmanship of the Administrator or his designee, consisting
of nine other members appointed by the President, including experts,
but none of whom are to be *federal* officers or employees. These can,
however, be employees of State governments or agencies, etc.

There is also an Effluent Standards and Water Quality Information
Advisory Committee,[150] consisting of a chairman and eight members
appointed by the Administrator, all being qualified to assess and evaluate
scientific and technical information. This Committee is enabled to
obtain information from federal agencies and is required in particular

to inform and advise the Administrator on any proposed draft regulations to be made under s. 304(b) of the Act. For this purpose, the Administrator must give the Committee full notice of the proposals 180 days before intended publication; and the Committee must provide such information as is available, together with their comments, within 120 days.

4 COMPARATIVE CONCLUSIONS

The US federal legislation sets out in great detail the required machinery for control of water pollution, and reserves to the Administrator of the Environmental Protection Agency (who is answerable to the Committee of Public Works and Transport of the House of Representatives for this purpose) the power to approve the standards laid down within each State – subject, of course, to the rights of a State under the Federal Constitution.

The approach adopted by the US legislation is to set scientific and technological standards in terms of 'national standards of performance', etc., and to relate these to the 'best practicable technology currently available', and/or 'the best available demonstrated control technology'. In the latter case, cost is taken into account, but to a limited extent.

However, in the UK, and in most other member States of the European Community, the broad principle normally applied is to prohibit all water pollution, but to provide for specific exceptions. Within this framework, however, the UK empowers regional Water Authorities – subject to ultimate control by the Secretary of State for the Environment – to exempt most types of discharge if they are not dangerously toxic. The advantage of this system (given that the officers concerned are incorruptible) is that reasonable allowance can be made for local conditions, including any intended industrial improvements and also the degree of local unemployment. The UK approach thus tends to be not unrelated to the long-established principle that the use of the 'best *practicable* means' of prevention may provide a defence in law. Nevertheless, this actual ground of defence is declining in significance, due to the increasing public attitude – which is reflected by the courts – that improvement and protection of the environment is of great importance, even in heavily industrialised localities.

Further, the UK approach places the burden of the initial assessment of a pollutant's characteristics upon the discharger, and requires that he finance any necessary means of monitoring and control. To some extent in contrast to this, it can, perhaps, be said that the Federal Government of the United States has very clearly provided, in the Clean Water Act, for a greater degree of federal aid for local disposal schemes, whilst, at the same time, specifically excluding provision of grants to industrial polluters.

This greater degree of central government assistance is also reflected in the specific provisions for funds to finance technical training and for more advanced scientific training of research; in addition, funds are provided for the setting up of both central and local research programmes into such matters as, for example, the effect of 'augmented stream flows', or as to systems of aquaculture, etc. In the UK, however, most research carried out by regional Water Authorities will be financed out of local water rates (i.e. tax levied upon water consumers) and land drainage rates. It may, however, be said that the US is better able to make provision for such funds, due to her far greater natural resources, including raw materials.

It is particularly significant that whereas on the setting up of regional Water Authorities, the UK disbanded the national Water Council (an action criticised by some experts at the time), the United States has provided, under the Clean Water Act, for both a Water Pollution Control Advisory Board, and (for more scientific requirements) an Effluent Standards and Water Quality Information Advisory Committee. The latter is specifically charged with the duty of advising the Administrator on any proposed draft regulations to be made in regard to 'national standards of performance' for the very wide list of types of source of pollution given in s. 306(b)(1)(A) of the Act. The Administrator can, in fact, prescribe the type of anti-pollution equipment to be used.

It may also be significant that, in the US, although proposals for means of control of discharges must be submitted by State Governors to the Administrator, a Governor is an elected individual; whereas the Water Authorities in the United Kingdom are non-elected bodies, appointed by central Government, and subject to the ultimate control of the Secretary of State for the Environment in England (or the Secretary of State for Scotland or for Wales), or by the Minister of Agriculture, Fisheries and Food. There is, thus, probably less localised pressure of public opinion upon a Water Authority; whereas, in the United States, the practical technical standards are geographically more remotely set by the Federal Government.

However, it may be as well finally to bear in mind a principle familiar to students of law that, to be fully effective, any system of law must take into account the characteristics of the persons to be controlled by it or otherwise concerned with it. In some countries local discretion may be safely and effectively permitted; whereas in others, a more strict system of policing may be preferable – combined, possibly, with less severe penalties.

NOTES

1. To adopt the description now favoured on the setting up of a fully elected European Parliament. The description, when used in this chapter, should be deemed to include all the previous separate European Communities, including the EEC. The member States of the Community are: Belgium, Denmark, Federal Republic of Germany, France, Ireland, Italy, Luxemburg, the Netherlands and the UK.

2. Austria, Belgium, Bulgaria, Canada, Cyprus, Czechoslovakia, Denmark, Finland, France, the German Democratic Republic, the Federal Republic of Germany, Greece, the Holy See, Hungary, Iceland, Ireland, Italy, Lichtenstein, Luxemburg, Malta, Monaco, the Netherlands, Norway, Poland, Portugal, Romania, San Marino, Spain, Sweden, Switzerland, Turkey, the Union of Soviet Socialist Republics, the UK, the USA and Yugoslavia.

3. The Democratic and Popular Republic of Algeria, the Arab Republic of Egypt, Israel, the Kingdom of Morocco, the Syrian Arab Republic and Tunisia.

4. D. Alastair Bigham, *The Law and Administration relating to Protection of the Environment*, (1973), and notably Supplement of 1975.

5. Public Health Act 1936, s. 92(1).

6. Ibid., s. 93.

7. Ibid., s. 94.

8. Ibid., s. 95.

9. Ibid., s. 96.

10. Ibid., s. 100.

11. Ibid., s. 259(1).

12. Ibid., ss. 261 to 265.

13. The Water Act 1973, s. 9.

14. Ibid., s. 11.

15. Ibid., s. 12.

16. Ibid., s. 13.

17. Ibid., s. 14.

18. Public Health Act 1936, s. 19.

19. Ibid., s. 27.

20. Water Act 1973, s. 14.

21. Ibid., s. 18.

22. Ibid., ss. 20, 21 and 23.

23. Ibid., s. 22.

24. Control of Pollution Act 1974, s. 31(1)(a).

25. Ibid., s. 31(1)(b).

26. Ibid., s. 31(1)(c).

27. Ibid., s. 31(1) & (2).

28. Ibid., s. 56(1).

29. Ibid., s. 31(1)(b).

30. Ibid., s. 56(1).

31. Ibid., s. 31(2)(a).

32. Ibid., s. 31(2)(b)(i).

33. Ibid., s. 31(2)(b)(ii).

34. Ibid., s. 31(2)(c).

35. Ibid., s. 31(2)(d).
36. Ibid., s. 31(2)(e).
37. Ibid., s. 31(3) to (6).
38. Ibid., s. 31(6) to (8).
39. Ibid., s. 32(1)(a)(i).
40. Ibid., s. 32(1)(a)(ii).
41. Ibid., s. 32(1)(a)(iii).
42. Ibid., s. 32(1)(b).
43. Ibid., s. 32(1)(c).
44. Ibid., s. 32(7).
45. As to which see Control of Pollution Act 1974, s. 33.
46. Control of Pollution Act 1974, s. 34.
47. Ibid., s. 34(1).
48. Ibid., s. 34(2).
49. Ibid., s. 36(1)(a).
50. Ibid., s. 36(1)(b).
51. Ibid., s. 36(1)(c).
52. Ibid., s. 36(6)(a), (b), (c).
53. Ibid., s. 35.
54. Ibid., ss. 38(1), 38(3)(a).
55. Ibid., s. 34(4)(a) to (g).
56. Ibid., s. 42.
57. Ibid., s. 43(3), (4), (5).
58. Public Law (P.L.) 94-83, Title: 'Background and Legislative History – The Issue'. (Note: the Section, subsection and paragraph numbers below can conveniently be referred to in the Committee Print combined text published by the US Government Printing Office in 1977, under Serial No. 95-12.)
59. P.L. 92-500.
60. P.L. 95-217.
61. S. 101(a).
62. S. 502(7).
63. S. 101(a)(6).
64. S. 101(d).
65. S. 102(a).
66. S. 102(b).
67. S. 102(b)(3).
68. S. 102(c) 2(A).
69. S. 102(c) 2(B).
70. S. 102(c) 2(C).
71. S. 104(a).
72. S. 104(a)(1) to (5).
73. S. 104(a)(6).
74. S. 104(b)(7).
75. S. 104(d).
76. S. 104(e) to (u).
77. S. 105.
78. S. 104(n)(2) & (4).
79. S. 106.
80. S. 111 and ss. 109 & 110.
81. S. 201.
82. Ss. 203 to 205.
83. S. 208.
84. S. 208(b)(1)(A) & (B).
85. S. 208(b)(2)(A) to (K).
86. As defined in s. 212(2).

87. S. 405(a).
88. S. 303 & 208(b)(4)(A).
89. S. 208(h).
90. S. 208(i).
91. S. 208(j).
92. S. 210.
93. S. 211.
94. S. 301(b)(1)(A)(i).
95. S. 301(b)(1)(A)(i) & (B); and s. 307.
96. S. 301(b)(2)(A).
97. Table 1 of Committee Print Numbered 95-30 of the Committee of Public Works & Transportation of the House of Representatives.
98. S. 301(c).
99. S. 301(g)(1)(C). Also note definition: carcinogenicity = cancer producing; mutagenicity = inducing unnatural mutants, i.e. changes; teratogenicity = producing monsters.
100. S. 301(h) & 402.
101. S. 311(a)(9) 'contiguous zone' means the entire zone established or to be established by the United States under Article 24 of the Convention on the Territorial Sea and the Contiguous Zone.
102. S. 302(a).
103. S. 302(b).
104. S. 303.
105. S. 303(a)(2).
106. S. 303(a)(2)(C).
107. S. 303(c).
108. S. 303(d)(1)(A).
109. S. 303(d)(1)(B).
110. S. 303(d)(1)(C) & (D).
111. S. 303(e)(3).
112. S. 212(2)(A) & (B).
113. S. 201.
114. But note the limitations (*ante*) as to grants for sewage collection systems.
115. S. 212(2)(A).
116. S. 212(2)(B). 'Combined sewers' are designed so as to be flushed by storm water, generally off road surfaces, etc.
117. S. 304.
118. S. 304(a)(1).
119. S. 304(f)(1), & (2)(A) to (F).
120. S. 304(g), (h) & (i).
121. S. 306(b)(1)(A):

> Pulp and paper mills; paperboard, builders paper & board mills; meat product and rendering processing; dairy product processing; grain mills; canned & preserved fruits & vegetables processing; canned & prepared seafood processing; sugar processing; textile mills; cement manufacturing; feedlots; electroplating; organic chemicals manufacturing; inorganic chemicals manufacturing; plastic & synthetic materials manufacturing; soap & detergent manufacturing; fertiliser manufacturing; petroleum refining; iron & steel manufacturing; non-ferrous metals manufacturing; phosphate manufacturing; steam electric power plants; ferroalloy manufacturing; leather tanning & finishing; glass & asbestos manufacturing; rubber processing, and timber products processing.

122. S. 509)b).
123. S. 307(a)(2).

124. S. 308(a)(4)(B).
125. S. 308(a)(4)(A).
126. S. 308(b).
127. S. 308(c).
128. S. 309(a)(1) & (b).
129. S. 309(c)(1). Fine of not less than $2,500, nor more than $25,000 *per day of violation.*
130. S. 309(a)(5)(B).
131. S. 309(e).
132. S. 309(a)(2) & (e).
133. S. 309(b)(1).
134. S. 311(a)(10).
135. S. 311(a)(2).
136. S. 311(b)(3).
137. S. 311(b)(4).
138. S. 311(b)(6).
139. S. 311(c)(1).
140. S. 311(c)(2).
141. S. 311(f)(2).
142. S. 311(f)(5).
143. S. 311(g).
144. S. 313(a).
145. Ss. 318 & 402.
146. S. 402(a)(1).
147. S. 404.
148. S. 501(e).
149. S. 503.
150. S. 515.

9 NUCLEAR POLLUTION OF THE MARINE ENVIRONMENT: SOME RECENT CONTROLS

Angelantonio D.M. Forte

1 PREFATORY REMARKS

This writer would submit, at the outset, that nuclear pollution, i.e. damage or harm to man or the environment[1] caused by radioactive matter, does not so much present the lawyer with problems concerning the formulation of principles of liability, rather it exercises his mind as to the form which possible controls, designed to minimise the occurrence of a recognised danger, should take. The general principles of law applied in the *Trail Smelter* case[2] could and probably would be applied today to any case involving pollution of the marine environment consequent upon a nuclear 'incident'.[3] Accepting, then, as it does, that pollution by nuclear material of the marine environment, so as to put others at risk in their legitimate use of ocean space, is contrary to international law, this chapter is concerned with some examples of the measures taken over the years by the international community to control the introduction into the seas of nuclear pollutants.

2 INTRODUCTION TO THE PROBLEM

Today, oil pollution of the world's oceans still perhaps constitutes the biggest single threat to that particular environment. There can be very few who have not heard of the massive damage inflicted upon not only marine life and vegetation but also on the communities who rely upon these, as a result of the wrecking of the *Torrey Canyon* and the *Amoco Cadiz*. Accordingly, international and municipal controls proliferate in an effort to minimise risk and regulate compensation in the event of pollution being caused by this source. There are, however, many other different kinds of pollutants which may be, and, indeed, have been, introduced into the oceans, and no one would deny that their impact on the marine ecosystem and, beyond that upon man himself, is potentially even more dangerous than any threat posed by oil spillage. Already, what many can merely imagine to be the effect of sea-borne radioactive contamination has been experienced as a reality by some people and countries:[4] the consequences of large-scale contamination from a nuclear source can only be guessed at with alarm.[5]

This essay is concerned with forms of nuclear pollution and yet it would be quite wrong to think that the nature of this problem is entirely *sui generis*, virtually demanding a *tabula rasa* approach to its solution. Useful perspectives may be gained from casting one's eye back to the problem of oil pollution in order, at least, to ascertain the major causes of that particular type of pollution and thereby identify similar areas of concern in the nuclear sphere. We know that oil pollution on a substantial scale commenced with the introduction of oil as a fuel source for shipping. We also know that oil pollution increased as the demands of industry necessitated the carriage of oil as cargo on tankers, and to state that the greater the demands of the developed nations, the greater the amount and risk of pollution, is merely to express the obvious. It would, however, be wrong to conclude that, despite the progression from VLCC to ULCC and despite disasters on the scale of the *Amoco Cadiz* stranding, the major source of oil pollution is accidental spillage. As has been recently shown, the figures demonstrate that the major source or cause of oil pollution is operational discharge.[6] Accidental spillages then, though they may assume spectacular forms and are no less disastrous for man and nature alike, are not as important a source of oil pollution as regular discharges into the sea of smaller amounts[7] by vessels in passage.

The lessons of the age of oil power are plain and its problems too will be paralleled as the world moves ever more rapidly into the nuclear power era. As resources of oil run down, nuclear energy will power the merchant fleets and navies of the world. As industries turn towards alternative energy sources, nuclear power will increase in use. With an increase in the use of nuclear energy there will come the problem of disposal of the detritus of the nuclear age. Radioactive waste and other materials will have to be transported and sea carriers will wish to increase their stake in the bulk handling of this dangerous cargo. This scenario is no idle projection for the future. The nuclear-powered ship is already with us, as is the carriage of radioactive material,[8] and these will surely increase in the future. With this increase in nuclear traffic at sea comes the risk of pollution. The nuclear ship will emit a certain amount of radioactive matter into the marine environment and a collision between two such vessels or vessels carrying a radioactive cargo may consign to the sea bed huge quantities of the most highly dangerous substances.

3 FORMS OR SOURCES

It may not be unfair to suggest that many people still perceive the problems and dangers of an era of nuclear energy as being essentially problems of the future.[9] Perhaps they do not think of pollution of the sea at all in this context. Nevertheless, the nuclear age has already made an impact on the world's seas and the forms of pollution already generated are worthy of discussion.

It is not difficult to identify the possible forms which nuclear pollution may take. They are, broadly speaking, as follows: (a) radioactive pollution caused by nuclear-powered vessels; (b) pollution caused by vessels carrying radioactive waste or other radioactive material; (c) pollution caused by the deliberate disposal at sea of radioactive matter; (d) pollution caused by the testing of nuclear devices; (e) contamination from land-based installations.[10]

It is to a consideration of the above pollution forms that we must now turn.

4 NUCLEAR SHIPPING AND CARRIAGE

There should, at once, be observed here a sharp dichotomy between the use of nuclear energy as a mode of propulsion for merchant vessels and for naval, i.e. military, vessels. The latter category greatly outnumbers the former and at present is made up largely of submarines.[11] Nuclear-powered, non-military vessels are not yet in common use,[12] but the basic problems to be faced by their operators are two in number, namely (a) access and (b) operational discharge.

4.1 Access

It is a basic principle of international law that there should be freedom of navigation on the high seas.[13] The extent of such waters is cut down by the exercise of sovereign rights by coastal States over an area of water adjacent to their coastlines termed the territorial sea.[14] Furthermore, although there exists a basic freedom of navigation right which is enjoyed by the vessels of other States through these territorial seas — i.e. a right of innocent passage[15] — that freedom is subject to restrictions imposed by the coastal State,[16] which may enact regulations to control, *inter alia*, health, security and navigation; it is also open to the particular coastal State to define a particular passage as not being innocent. Furthermore, in a zone of the high seas, contiguous to its territorial sea, a coastal State may also take necessary steps to prevent infringements of health regulations in operation in its territory or territorial seas.[17] These are mentioned because at present many States[18] might be reluctant to permit access to their ports or passage through their territorial waters of nuclear vessels. Indeed, so concerned have some States become about the impact of pollutants upon their off-shore resources that they have taken unilateral action to establish anti-pollution zones; for example, in

1970 Canada passed the Arctic Waters Prevention of Pollution Act which created an anti-pollution zone of 100 miles. The UK, whilst stopping short of declaring a 'Canadian type' zone, has also taken steps under the Prevention of Oil Pollution Act 1971, to take any necessary action against foreign ships on the high seas to protect UK territory or territorial waters from oil pollution.

The UK has twice had to deal with proposed voyages to its territories of a non-military nuclear ship, the *n.s. Savannah*. Its method of dealing with the problem of access to its ports through territorial waters was to enter into a bilateral agreement with the US specifying the precise terms and conditions to which rights of access and passage were to be subject.[19] It would be tedious to repeat here the full terms of these Agreements. However, a brief look at three of them may suffice to indicate the sort of controls which States may wish to exercise in the future.[20]

(i)   Access was conditional upon the *Savannah* complying with international requirements concerning the safety of life at sea.[21] Clearly one would expect a nuclear vessel to measure up to the same high standards of safety imposed on vessels powered by more conventional means.

(ii)   Entry to the *Savannah* was to be freely available for the purpose of ensuring that she was in safe operational condition. Under UK domestic legislation, shipowners are obliged to ensure that their vessels are seaworthy,[22] but all ships in a British port, whether British or foreign-owned, may be challenged on grounds of safety.[23] Any vessel suspected of being unsafe may be detained and subsequently surveyed and, if found to be in an unsafe condition, detention will be continued until such time as the vessel is made safe. *A fortiori*, the nuclear ship should be subject to no less a control.

(iii)   Disposal of the *Savannah's* radioactive waste was to be made through the agency of the appropriate domestic authorities. No coastal State would wish uncontrolled discharge of such matter into its territorial waters.

4.2 Operational Discharge

Intimately connected with the question of access is the fear of coastal States of the effects of any operational discharges of radioactive matter whilst the vessel is passing through their territorial seas and adjacent waters. The International Convention for the Prevention of Pollution by Ships 1973 (MARPOL) tackles on a comprehensive basis all types of operational discharge, but not the discharge of radioactive matter. This is dealt with by the International Convention for the Safety of Life at Sea 1960[24] (SOLAS 1960) which, amongst other things, covers the safety of passengers, crew members and marine resources from radioactive contamination by nuclear ships.

The other source of pollution which requires to be discussed here is accidental pollution. All of the controls designed to minimise the risk of accidental nuclear pollution and to provide compensation for damage caused thereby have had to balance requirements of stringent security with the beneficial development of the nuclear industry. In other words, compensation should be designed to be prompt and sufficient and yet, at the same time, it should not be prohibitively high and thus reduce incentive. This has been achieved by a number of Conventions, all of which combine the principle of strict liability[25] with a ceiling on the amount recoverable and a limitation of actions period.

There are a number of Conventions which deal with liability for nuclear accidents, though not all of these deal with accidents involving nuclear-powered ships.[26] This is rectified to some extent by the International Convention on the Liability of Operators of Nuclear Ships 1962, which was passed in Brussels. The Brussels Convention provides that the operator of a nuclear vessel shall be strictly liable for damage caused by a nuclear accident up to a maximum of 1,500 million francs; there is a ten-year limitation period for claims. Since, however, the Convention seeks to deal equally with military and other State-operated vessels as well as with non-military vessels, the Brussels Convention has yet to be ratified by States which possess and operate nuclear-powered ships.

The carriage by sea of nuclear material as cargo is dealt with by both SOLAS 1960[27] and the Brussels Convention relating to Civil Liability in the Field of Maritime Carriage of Nuclear Material 1971. SOLAS 1960 lists radioactive material as one of its several categories of dangerous goods. Accordingly, there are prescribed extremely detailed conditions as to documentation, labelling, packing and storage.[28] In addition, IMCO has played a role here and in 1966 it adopted an International Maritime Dangerous Goods Code which included radioactive cargoes. The Brussels Convention was designed to avoid anomalies caused by overlapping provisions relating to liability and its main thrust is to exclude a carrier from liability under any other provision relating to damage caused by radioactive substances if, under either the Paris or Vienna[29] Conventions, the operator of the nuclear installation is liable.

5 DISPOSAL OF NUCLEAR MATERIALS

The problem considered here is the dumping at sea of containerised radioactive waste, the by-product of nuclear industry. This is usually done far out at sea in international waters and may well constitute one of the most invidious forms of pollution of the ocean depths. The corrosive and destructive power of the sea cannot be underestimated and no one can predict that the materials[30] used to construct these containers will do any more than slow down the rate of dispersal of their contents. Hopefully, the dispersal rate will be sufficiently low to preclude the risk of contamination, especially with the growth in recent years of nuclear industries.

The Stockholm Conference on the Human Environment of 1972 stated as a basic principle that States should take every possible measure to prevent pollution of the seas by dumping hazardous matter therein.[31] Pursuant to this there were concluded in 1972 two such Conventions, viz.: the Convention on the Prevention of Marine Pollution by Dumping of Wastes and Other Matter (the London Convention); and the Convention for the Prevention of Marine Pollution by Dumping from Ships and Aircraft (the Oslo Convention).

The London Convention is a global Convention and prohibits the dumping of high-level radioactive waste;[32] low-level nuclear wastes may be dumped, but only under permit issued by a competent national authority. Naval and other government vessels are not included in the Convention, although each signatory State must adopt measures in respect of such vessels consistent with the purpose and object of the Convention. The parties to the Convention are not only obliged to issue permits allowing dumping, but must also maintain records and monitor the effects of such activities. Provision is also made to promote regional agreements to control pollution. Under the UK legislation, warships are subject to control, and the responsibility for issuing permits is in the hands of the Minister of Agriculture, Fisheries and Food. In Scotland, responsibility for dumping within territorial waters lies with the Secretary of State.

The UK has given effect to the London and Oslo Conventions by passing the Dumping at Sea Act 1974.

The Oslo Convention does not deal with radioactive materials, and is an example of control of pollution on a regional basis, applying as it does to only the North-East Atlantic. Some other examples of regional controls may be found in the Antartica Treaty of 1959 and the Convention on the Protection of the Marine Environment of the Baltic Sea Area 1974. Both of these measures deal with the dumping of nuclear waste, which is generally prohibited by their terms.

6 TESTING OF NUCLEAR DEVICES

In July 1946 the US commenced the testing of nuclear weapons on
Bikini Atoll in the Pacific Ocean. At Bikini and other sites in the
Pacific subsequently chosen for the purpose, the practice was adopted
of warning shipping to avoid a specified area of the surrounding waters.[33]
However, on one occasion in 1954 a disastrously inaccurate calculation
of the danger zone resulted in 82 Marshall Islanders being injured. Moreover, the crew of a Japanese fishing vessel some fourteen miles beyond
the warning zone were also injured and one man died. A large quantity
of tuna was subsequently condemned as being unfit for consumption.
The US made an *ex gratia* payment of two million dollars to Japan, but
it did not admit liability.

Commentators were divided on the legality of the testing of nuclear
weapons. McDougal and Schlei[34] justified the tests on the ground that
they did not violate the principle of freedom of the seas. Others, however, regarded such testing very differently, arguing that the 'fencing-off'
from the maritime and air traffic of other nations of hundreds of
thousands of square miles of open sea and air space was contrary to the
principle of freedom of the seas.[35] They added that freedom to fish
had been violated by excluding fishing boats from catching in the areas
adjacent to the warning zones.[36]

In the event, it seemed as though the pro-testing lobby won the day
and a Soviet bloc proposal at UNCLOS I, designed to prohibit the testing
of nuclear devices on the High Seas, was never implemented. However,
since the uncontrolled testing of the fifties, events have taken a different
turn with the coming into force of the 1963 Test Ban Treaty which
prohibited the testing of nuclear weapons in the atmosphere, in space
and under water. One of the major drawbacks, however, was the fact
that not all States possessing a nuclear capability were party to the
Treaty, China and France being significant exceptions.

France recently tested nuclear devices in the Pacific and quickly
discovered the legality of its actions challenged in the International
Court of Justice by Australia and New Zealand.[37] The basic issue, which
was surely the prevention of any future atmospheric testing, was in fact
neatly side-stepped by the Court turning its ear to French representations
that the testing would in fact now cease. Thus the question of legality
was not really considered. It might be the case that this issue would
require to be resolved on a consideration of Art. 2, para. 2, of the
Geneva Convention dealing with reasonable use of the high seas. If that

is correct, then, despite the seeming legality of the earlier tests carried out by the US, USSR and UK, it might well be the case now that changes in world opinion, coupled with the current practice of underground explosions by many States, would justify the conclusion that such tests are illegal.[38]

The 1971 Treaty on the Prohibition of the Emplacement of Nuclear Weapons and Other Weapons of Mass Destruction on the Seabed and Ocean Floor and in the Subsoil precludes the placing of, *inter alia*, nuclear weapons on the sea bed and its subsoil.[39]

7 LAND-BASED CONTAMINATION

At an international level, the control of land-based pollution from nuclear sources has only been dealt with by regional agreements. The North-East Atlantic area is now covered by the Paris Convention of 1974.[40] Predicated on the 1972 Oslo Convention, it, too, utilises the idea of 'black' and 'grey' listing of pollutants adopted in the Conventions on dumping at sea. Radioactive materials were omitted from the list of proscribed substances under the Oslo Convention. However, they do fall within the controls imposed by the Paris Convention, the Contracting Parties being obliged to take steps to prevent and eliminate radioactive pollution of the marine environment from land-based sources.[41]

8 CONCLUSION

At present the framework of controls designed to meet the problems of the nuclear age present a complex and somewhat fragmented pattern. In some instances control is attempted at a global level; in others, regional agreements have been achieved; and at national levels action, indeed sometimes unilateral action, has been taken.

The picture, then, for the future is certainly not discouraging on the whole, and possibly justifies a feeling of cautious optimism. A problem, however, may be that of a clash or conflict between the various levels of control. On balance, one conceives the role of the global Convention as most appropriate to the control of nuclear shipping, the carriage of radioactive freight and the general problem of pollution of the high seas. It is at the regional level, in particular, that one finds an almost natural forum for the discussion of issues such as dumping and discharge from on-shore facilities. In the interests of homogeneity, however, regional controls should always reflect or operate within the general principles of the global schemes. This task is far from finished, but an encouraging start has been made.

NOTES

1. A precise definition of the term 'pollution' is, as has been rightly pointed out, extremely difficult. See Springer, 'Towards a Meaningful Concept of Pollution in International Law', [1977] 26 *I.C.L.Q.*, 531-57.

2. 3 R.I.A.A. 1905.

3. The point is most cogently argued by Professor Fleischer in 'Pollution from Seaborne Sources', pp. 78-102 of Vol. III of *New Directions in the Law of the Sea*. *Rylands* v. *Fletcher* (1868) L.R. 3 H.L. 330 would be the common law paradigm in Anglo-American legal systems.

4. Eaters of laverbread (a seaweed delicacy) in Cumbria and Wales have been exposed to doses of radiation from eating plants which were gathered near Windscale on the Cumbrian coast. See the *Fourth Report of the Royal Commission on Environmental Pollution*, Cmnd. 5780. See also the Japanese incident discussed in the text.

5. Scientific monitoring of pollution is at least as important as the devising of legal controls. See Schachter and Stewart, 'Marine Pollution Problems and Remedies', [1971] 65 *A.J.I.L.* 84-111.

6. Abecassis, *The Law and Practice Relating to Oil Pollution from Ships*, Table 1.1; and Table 1 in Ch. 1, *supra*.

7. Though size is surely relative. At the time of writing this chapter a twelve-mile long oil slick in the Firth of Forth has taken a high toll of gannets and other diving birds. Many tankers now lie in the Firth prior to discharge in Rotterdam.

8. Already material is being transported from Japan to the UK for reprocessing.

9. Though the Harrisburg near 'melt-down' may now have heightened the consciousness of ordinary people and not just pressure groups.

10. Consideration of the floating nuclear plant (f.n.p.) would perhaps be too premature here. The reader is directed to Blake, 'Floating Nuclear Plants – A "Reasonable Use" of the High Seas?', [1978] *California Western Internat. Law. Jo.*, 191-227.

11. See generally Janis, *Sea Power and the Law of the Sea* (1976). The US has at present some four dozen such vessels; the Soviet Union has a similar number.

12. The best-known such vessel is the United States' *n.s. Savannah*. The Federal Republic of Germany has the *n.s. Otto Hahn*.

13. Art. 2(1), Convention on the High Seas of 1958.

14. Art. 1, Convention on the Territorial Sea and the Contiguous Zone of 1958.

15. Ibid., Arts. 14-23.

16. Ibid., Art. 17.

17. Ibid., Art. 24.

18. E.g. Japan.

19. Exchange of Notes relating to Use of United Kingdom Ports and Territorial Waters by the *n.s. Savannah*, London, 19 June 1964: Exchange of Notes modifying the Agreement of 19 June 1964, relating to Visit of *n.s. Savannah* to Hong Kong, London, 12 June 1967.

20. See App. 1 of the 1964 Exchange for the full conditions.

21. The requirements specified were the International Regulations for Preventing Collisions at Sea; the International Convention for the Safety of Life at Sea.

22. The 1949 Act, s. 29.

23. The 1894 Act, s. 459 *et seq*.

24. A new Convention was adopted in 1974. This will replace the 1960 Convention, as amended, though it reproduces the earlier provisions concerning nuclear vessels.

25. Thus eliminating the need to prove negligence.

26. Thus neither the Convention on Third Party Liability in the field of Nuclear Energy, 1960 (the Paris Convention) nor the Convention Supplementary to the Paris Convention on Third Party Liability in the Field of Nuclear Energy, 1963, deals with this issue.

27. Once more the 1974 Convention merely reproduces the provisions of the earlier Convention.

28. In the UK the position is governed by Rules made under s. 23 of the 1949 Act which include radioactive substances.

29. The Convention on Civil Liability for Nuclear Damage, 1963. This is basically an attempt to extend the Paris Convention on a global basis.

30. Concrete and steel.

31. Princ. 7. See also Arts. 2 and 25 of the Geneva Convention on the High Seas 1958.

32. Some relaxation is allowed in the case of emergencies.

33. E.g. in the case of the March 1954 test at Enewitok Atoll the area was 400,000 square miles. See McDougal and Schlei, 'The Hydrogen Bomb Tests in Perspective: Lawful Measures for Security', (1954-5) 64 *Yale L.J.* 648.

34. *Loc. cit.*

35. Art. 2, Geneva Convention on the High Seas 1958.

36. See, for example, Margolis, 'The Hydrogen Bomb Experiments and International Law', (1954-5) 64 *Yale L.J.* 629.

37. These are the so-called *Nuclear Tests Cases* (*Australia* v. *France*; *New Zealand* v. *France*), 1974 I.C.J. Rep. 253 and 457.

38. Tiewel, 'International Law and Nuclear Test Explosions on the High Seas', 8 *Cornell Internat. L.J.* 45. This writer concludes that nuclear testing does constitute an unreasonable use of the high seas.

39. There is a third Convention, namely the Convention on the Prohibition of the Development, Production and Stockpiling of Bacteriological and Toxic Weapons and on their Destruction 1972.

40. Convention for the Prevention of Marine Pollution from Land-Based Sources 1974.

41. Under the 1974 Baltic Convention radioactive substances may not be introduced into that area without a special licence.

Part III:
COMPARISONS

10 LAND-BASED MARINE POLLUTION IN THE EUROPEAN COMMUNITY

D. Alastair Bigham

1 INTRODUCTION

As mentioned in the introduction to Chapter 8, the intention in this present chapter is to consider the legislative provisions made by the European Community in seeking to remedy the problem of water pollution – and, specifically, land-based marine pollution.

In providing for control of pollution of various types (and initially, water pollution, atmospheric pollution and noise), it has been necessary for the Community to take into account the sometimes quite marked differences between the national economies of member States, in terms of the extent to which each State may rely primarily upon manufacturing industry, agriculture or tourism. In addition, it is a relevant factor that there is, by comparison to many other areas of the world, a relatively high density of population with a desire for a relatively high standard of living.

Before embarking upon discussion of this legislation, it is advisable to refer briefly to certain aspects of the organisation of the European Community, and also to distinguish between the several types of legislation of which the Community, as such, is capable.

The principal governmental and legislative body of the European Community is the Council of Ministers, which consists of representatives of each and every member State, appointed (in a sense) as delegates by the State governments. For most purposes the votes of each of the member States within the Council are weighted and, upon that weighting, a 'qualified majority' of 41 votes is normally required to pass legislation. The weighting for qualified majority votes is given in footnote 1.

In addition to the Council of Ministers, the government of the Community includes the European Parliament – which is at present essentially a debating chamber which is of great value in sounding opinion, but is not a legislative body; and the Commission of the European Community or Communities (the European Commission). The Commission is essential to the progress of the Community. The European Commissioners as individuals are each responsible for specific areas of European government, and they report direct to the Council of Ministers with their proposals. Although member States, through their Members of the Council, can also put forward proposals, these will normally be investigated by the Commission, which will then report back to the Council.

Clearly the above description, due to brevity, is not fully accurate, but it suffices in terms of explaining the types of legislation which will be referred to below. These are Council Decisions, Council Directives, Regulations and Council Opinions or Declarations. Opinions or Declarations are not, in strict terms, legislation, but they will always be strongly persuasive.

A Council Decision is full and complete legislation, in that it leaves no discretion as to the means of application within each member State. This should be distinguished from a Council Directive, which has the effect of applying a firm requirement that all member States must accept the objectives given in the Directive, but leaves the means (except where otherwise detailed) to the individual member States. Regulations are made by the Commission, and relate to matters which have either been fully agreed by the Council in terms of the mode of the application which will not cause conflict of views between member States; or, perhaps, are of lesser importance in terms of principle: they are, however, of *immediate and totally binding effect* throughout the Community.

2 LEGISLATIVE PROVISIONS AND PROPOSALS BY THE EUROPEAN COMMUNITY

As early as 1963 it was realised that the severity of the pollution of the River Rhine was such as to necessitate an international agreement, and the Agreement concerning the International Commission for the Protection of the Rhine against Pollution was signed in Berne.[2] The measures then taken have not, however, been sufficient. The Rhine still carries much industrial waste down river from Switzerland, Germany and France, and has been said to be 'virtually an open sewer' by the time it reaches the Netherlands. However, more recently, severe penalties for water pollution have been imposed by these signatory countries. The need for international or Community co-operation can be plainly seen if it is realised that over 80 per cent of European lakes and rivers are shared by two or more countries[3] (not all of which are member States of the Community).

The urgent need for mankind to take steps to protect the environment was first strongly emphasised on a world-wide level at the United Nations Conference on the Human Environment held in Stockholm in

June 1972. The European Community formally acknowledged this need at an EEC summit conference in Paris on 19 and 20 October 1972, at which it was decided that a programme of action based upon a precise timetable should be established not later than 31 July 1973. This conference was followed later in the same month (31 October) by a meeting of the Environment Ministers of the member States, held in Bonn.

The process of adoption and implementation of positive steps to protect and improve the environment in the Community has suffered delay, but the *First Action Programme on the Environment* was adopted by the Council of Ministers in a Declaration of the 22 November 1973.[4] As already indicated, Council Declaration does not, in fact, constitute legislation, but it does represent a firm statement of Community policy. The objectives of this policy have been concisely stated as follows:

> The aim of a Community environment policy is to improve the setting and quality of life, and the surroundings and living conditions of the peoples of the Community. It must help to bring expansion into the service of man by procuring for him an environment providing the best conditions of life, and reconcile this expansion with the increasingly imperative need to preserve the natural environment. It should:
> (i) prevent, reduce and as far as possible eliminate pollution and nuisances;
> (ii) maintain a satisfactory ecological balance and ensure the protection of the biosphere;
> (iii) ensure the sound management of and avoid any exploitation of resources or of nature which cause significant damage to the ecological balance;
> (iv) guide development in accordance with quality requirements, especially by improving working conditions and the settings of life;
> (v) ensure that more account is taken of environmental aspects in town planning and land use;
> (vi) seek common solutions to environmental problems with States outside the Community, particularly in international organizations.[5]

At the request of the Council, these objectives are being further pursued and extended in a *Second Action Programme on the Environment* for the period 1977 to 1981,[6] which includes the principle that:

> This work is aimed at developing a method, relying chiefly on
> cartography, which will enable environment-related data and values
> to be introduced into the physical organization of space usage and
> provide a means of matching economic and social demand to
> ecological supplies.[7]

Combined with this general strategy, a further essential principle has
been adopted, and stated thus:

> The best environment policy consists in preventing the creation of
> pollution or nuisances at source, rather than subsequently trying to
> counteract their effects. To this end, technical progress must be
> conceived and directed so as to take into account the concern for
> protection of the environment and for the improvement of the quality
> of life, at the lowest cost to the Community. This environment policy
> can and must go hand in hand with economic and social development,
> and also with technical progress.[8]

> The cost of preventing and eliminating nuisances, as a matter of
> principle, be borne by the polluter.[9]

The Second Action Programme adopts certain approaches which
should be borne in mind when considering individual items of legislation.
Title I of the programme re-states the objectives given in the First Action
Programme without deviation from them. Title II is concerned with
reduction of pollution and nuisances, and develops further ideas in
regard to water pollution (and also as to atmospheric pollution and
noise). Title III is concerned with achieving the non-damaging use and
effective management of land, including the protection of flora and
fauna and of natural resources generally, by means of an ecological
mapping system. This general protection also involves the conservation
of resources and the reduction of wastage. Title IV introduces the
principle of a general and positive programme to protect and improve
the environment, the mechanics of which include the concept of
Environmental Impact Assessment (E.I.A.) (*post*), in which State and
local authorities and industrialists, etc., will have positive responsibilities.

Within the broad policies stated in the Second Action Programme,
the 'polluter pays' principle is established, on the basis that subsidies
would distort genuine competition in the European Common Market

(the 'market philosophy'). And, in order to put the 'polluter pays' principle into effect, a Council Recommendation[10] provides by definition for environmental quality standards, products standards, process standards, emission standards, installation design standards and operating standards. However, although it is helpful to know that these standards exist, it is not possible to discuss them further in this chapter.

In support of the Second Action Programme, the Community also set up a research programme[11] for the period 1976-80, to develop scientific knowledge and practical application in regard to four specific areas of the environmental action programme, namely: to establish criteria in relation to acceptable exposure and effects in regard to pollutants and potentially toxic chemicals; to organise the management of environmental information, in particular in regard to chemicals; to consider means of reducing and preventing pollution and nuisances, and the application of 'clean' technologies; and to study means generally for the protection and improvement of the natural environment. In this research the Community institutions co-operate with other scientific bodies, such as the Committee on Scientific and Technical Research (CREST); the Scientific and Technical Information Committee (STIDC) and Euronet, in order to avoid duplication of effort. Euronet is a transmission system for data from more than twenty computer banks, which will provide answers within three minutes. It offers four communications centres, namely London, Paris, Rome and Frankfurt, with 'concentrators' in Dublin, Brussels, Copenhagen and Amsterdam; and local access in Luxemburg. The transmission operates through DIANE (Direct Information Access Network for Europe).

By a Council Decision of 8 December 1975,[12] a procedure to enable the Commission to maintain a comprehensive inventory of sources of information on the environment within the Community was introduced, and such information continues to be recorded on an annual basis.

It is inappropriate in this discussion limited to marine pollution to refer to other more general Community arrangements, save to say that there is positive co-operation between the European Community and other organisations, such as the United Nations Environment Programme (UNEP), the Inter-Governmental Maritime Consultative Organisation (IMCO) and the Organisation for Economic Co-operation and Development (O.E.C.D.) Environment Committee.

To summarise the objectives now firmly adopted by the Community, these can be briefly stated as follows:

(i) the best environment policy consists in preventing the creation

of pollution or nuisances at source rather than subsequently
trying to counter their effects;

(ii)   environment policy can and must be compatible with economic
and social development;

(iii)   effect on the environment should be taken into account at the
earliest possible stage in all technical planning and decision-
making processes;

(iv)   any exploitation of natural resources or anything which causes
significant damage to the ecological balance must be avoided;

(v)   standards of scientific and technological knowledge in the Community
should be improved with a view to taking effective action to conserve
and improve the environment and to combat pollution and nuisances.
Research in this field should therefore be encouraged;

(vi)   the cost of preventing and eliminating nuisances must in principle
be borne by the polluter;

(vii)   care should be taken to ensure that activities carried out in one State
do not cause any degradation of the environment in another State;

(viii) the Community and its Member States must take into account
in their environment policy the interests of the developing
countries, and must in particular examine any repercussions
of the measures contemplated under that policy on the
economic development of such countries;

(ix)   the Community and the Member States must make their
voices heard in international organizations dealing with aspects
of the environment and must make an original contribution to
these organizations;

(x)   the protection of the environment is a matter for all in the
Community, who should therefore be made aware of its
importance;

(xi)   in each different category of pollution, it is necessary to
establish the level of action that befits the type of pollution;

(xii)   major aspects of environmental policy in individual countries
must no longer be planned and implemented in isolation;

(xiii) Community environment policy is aimed, as far as possible, at
the co-ordinated and harmonized progress of national policies
without, however, hampering potential or actual progress at the
national level. However, the latter should be carried out in a
way that does not jeopardize the satisfactory operation of
the common market.[13]

2.1 The European Community's 'Coastal Area'

Before referring in more detail to the individual Council Decisions and Directives already indicated, it may be helpful to consider briefly certain provisional ideas which are likely to be put forward by the European Commission in regard to the 'space usage' (and *aménagement du territoire*/town and country planning) of the coastal areas of the Community, since these give a clear picture of some of the major problems involved in regard to pollution of the seas from land-based sources.

Within the Ecological Mapping Project, a *'coastal area'* for the entire coastline of the European Community (including islands) is visualised, and this is the subject of a special study as a result of the severe problems which apply to it, due to a rapidly deteriorating environmental situation.

Any defects in the planning and management of the area must have some bearing, direct or indirect, upon land-based marine pollution. Even the presence of large numbers of holiday-makers will affect the immediate coastal waters as a result, for example, of increases in sewage sludge, or even in the use of petrol driven boats in significant numbers. These influences have some effect upon the balance of the marine ecosystem. Where coastal developments include local manufacturing or processing plants, there are substantially greater problems in connection with waste disposal.

The initial, tentative proposals now being considered by the European Commission[14] are based upon pilot surveys of the coastal area of Brittany (France) and of Puglia (Southern Italy). They indicate that immediate steps should be taken to protect the European coastal area, both in terms of the use and planning of coastal land, and the use and exploitation of the sea and sea bed. This requires a concerted approach within the Community so as to alter certain of the more recent trends in use of land, and to provide improved physical and local economic planning for the future. The 'coastal area' requiring protection has been indicated as, in general, a coastal landward strip of 10km, but variable according to the existence of such factors as urban developments in the form of expansion of ports inland, or of physical factors which restrict the width of relevant coastal land, such as high mountains. The 'coastal area' should also include the sea to, in most cases, the extent of the continental shelf, but with variations for such factors as unusually shallow waters, or the existence of large numbers of islands.

Thus it appears to be the present intention that, at such time as the European Community positively defines its coastal area, provision will be made for improved *aménagement du territoire*, which will in itself improve the means of control of land-based marine pollution. It is

provisionally proposed that such measures should include specific
attention to marine ecosystems, more particularly in regard to the
husbanding of natural resources and fishing reserves.

The aims to be adopted in due course are likely to include the more
careful siting of heavy industry and industry having the greatest pollution
risk, while favouring 'clean' and light industry, including the encouragement of local handcrafts where appropriate. At the same time, it is
probable that exploitation of the sea bed will be more strictly controlled.

The draft report also visualises the setting up of research institutes
and observation posts; and the use of aerial photography. These would
have the effect of encouraging greater co-ordination between maritime
authorities, land authorities (both national and local), joint maritime
and land authorities (such as port authorities) firms operating off-shore,
firms operating on land, and bodies representing, variously, coastal
landowners, conservationists and the local population at large.

The funding of the necessary steps required to protect the coastal
area would involve the European Community as a whole, in addition to
a degree of financial support at national levels.

Specific attention has been drawn in the draft study to the means of
developing improved administrative machinery for effective environmental control, and the creation or improvement of development plans
at both national, regional and local levels. Within land-use development
plans, there should particularly be clear indications of locations suitable
for production activities and processing activities; and there should also
be provision for the implementation of a clean water policy which meets
the important need to protect and to keep checks on marine wetlands
at river mouths, whilst maintaining the quality of water supplies to
built-up areas on the coast.

It is significant that this approach to local planning and water pollution
control includes the possibility of eventual legislation, requiring that
the principle of Environmental Impact Analysis (E.I.A.) should apply to
all significant industrial and other developments. Discussion of the
various approaches to legislation requiring the use by developers, or by
'responsible authorities' (to use the Community's technical expression),
of Environmental Impact Statements (E.I.S.) is, unfortunately (although
indirectly very relevant), beyond the space available here. However, it
may be sufficient to say that industrial and other developers contemplating the siting or expansion of manufacturing or processing plants, both
in the coastal area and elsewhere in the European Community, may be
required in the near future to produce an E.I.S. or E.I.A., which will be
either (if so provided in Community legislation) at their own expense, or

with the cost shared with the responsible planning authority. Such a study will involve a careful survey of the environmental and other effects of the proposed development; and, in the case of an Environmental Impact Analysis, will involve the provision of quite detailed information about the possible and probable effects of the development upon local ecosystems. In these respects, of course, the observation posts envisaged for monitoring purposes, and the related research institutes, would be of great value so far as sites in the coastal area are concerned. For present purposes, the writer would merely observe that if such a system of Environmental Impact Analysis is eventually adopted, the local planning authority should be required to share half of the cost of the survey, specifically because (a) as a permanent local organisation, an authority is in a position to monitor and record such matters on a continuing basis; and (b), of equal importance, a requirement that the local authority should meet half the cost of such an analysis ought to be persuasive in keeping down the authority's expenditure involved in such research and recording.*

In terms of the effective application by individual member States of the aims and objects of any future legislation, it has been suggested that there should be integrated coastal planning between State frontier regions, and also in regard to development of specific measures to combat the major pollution risks. In addition, it has been suggested that a common organisation should be set up 'to supervise economic activities carried out at sea'. The extent and form of these fields of supervision have not yet been clarified; but, no doubt, such factors as the degree of exploitation of fisheries, or of the sea bed, could be rendered more appropriate to overall environmental requirements by means of financial adjustments within the Common Market economic system.

Finally, as to provisions for protection of the coastal area of the European Community, it should be observed that the draft study suggests the need for co-ordinated effort between groups of member States having, respectively, coastlines on the Baltic Sea, the North Sea, the Atlantic Ocean and English Channel and the Mediterranean Sea.

The general coastal area problem includes the need to take into account the Directives already adopted by the Community which are specifically concerned with control of pollution of both fresh water and sea water. These Directives, as has already been indicated, largely relate to several international Conventions, and in particular to those signed at Oslo,[15] London,[16] Helsinki,[17] Paris[18] and Barcelona.[19] But it is also necessary to take into account further international agreements and organisations (which are discussed in other chapters in this book), in

*See pp. 287-9 for an up-to-date version of this paragraph.

regard to navigation, discharges of oil from ships and off-shore oil platforms; and fishing. In this latter area, it is likely to be the policy of the European Community to seek to persuade all member States to adopt essential safety rules as to navigation and as to the operation of off-shore platforms, in accordance with the Council Resolution for the Purpose of Reducing Pollution caused by Oil Spills at Sea of 30 May 1978.

In addition, the Community may seek to encourage increased co-operation between the existing European maritime and fisheries organisations; between member States as to shellfish breeding waters; and by increased co-operation under the several relevant international Conventions.

2.2 European Community Provision as to the Various Types and Sources of Water Pollution

Reference has already been made to the First and Second Action Programmes on the Environment, and it is significant that in 1973 the First Programme described sea water as:

an essential source of products or proteins, which are extremely valuable in a world which is becoming increasingly overpopulated. In addition, the sea plays a vital role in maintaining the ecological balance by supplying a large proportion of the oxygen upon which life depends. The sea and coastal areas are also of tremendous importance for recreation and leisure.

The pollution of the sea has already reached a high level. For example, a disturbing accumulation of certain pollutants can be detected in plankton, other living organisms and sediments, and even now there is evidence of a not insignificant danger of eutrophication[20] in certain estuaries and coastal areas.[21]

Following the Community's acceptance of this scientific and economic statement, several international conventions have been signed and ratified by the Community and by individual member States; and, in consequence or in addition, a number of important Council Decisions and Directives, together with certain Regulations, have been issued as legislation. These will be discussed in the following pages, and footnotes as to official reports and references are provided. These legislative provisions are given below.

*2.2.1 The Paris Convention for the Prevention of Marine Pollution from
Land-based Sources, 4 June 1974.*

The Paris Convention was signed on 4 June 1974 by 14 Contracting
Parties; namely, 8 member States of the European Community and the
Community itself (as an individual signatory), and 5 other sovereign
States.[22] The member States were Belgium, Denmark, the Federal
Republic of Germany, France, Ireland, Luxemburg, the Netherlands and
the United Kingdom. The other five Contracting Parties were Iceland,
Norway, Portugal, Spain and Sweden.

So far as the European Community was concerned (as an individual
contracting party) the matter was concluded by ratification as a result
of two Council Decisions of 3 March 1975;[23] and, following ratification
by 7 of the Contracting Parties, the Convention came into effect on
6 May 1978. However, as at March 1979, 6 out of the 14 signatories
have not yet ratified,[24] although it appears that they fully intend to do
so, once various problems of detail affecting national legislation have
been resolved. (In this respect, it is significant that those of the leading
signatories in the economic sense which have not yet ratified have in
fact been taking an active part in the work of the Interim Commission
and Commission set up under the Convention.) In view of the strong
probability that all signatories will have ratified the Convention in the
very near future, it is not intended to discuss here those areas of national
laws which differ from the objectives laid down in the Convention.

In the Paris Convention, the Contracting Parties pledged themselves:

> to take all possible steps to prevent pollution of the sea, by which is
> meant the introduction by man, directly or indirectly, of substances
> or energy into the marine environment (including estuaries) resulting
> in such deleterious effects as hazards to human health, harm to living
> resources and to marine ecosystems, damage to amenities or interfer-
> ence with other legitimate uses of the sea.[25]

(It should be observed that the reference to 'interference with other
legitimate uses of the sea' appears to introduce an extension to the
more conventional type of definition of 'pollution', and may tend to
produce problems of legal and practical interpretation in certain factual
situations.)

However, this undertaking does not comprise the entirety of the
definition of marine pollution from land-based sources adopted in the
Convention, and it is necessary also to refer to Article 3 of the Convention, which states that it means:

the pollution of the maritime area (*post*)
  (i)   through watersources,
  (ii)   from the coast, including introduction through underwater
         or other pipelines,
  (iii) from made-made structures placed under the jurisdiction of a
         Contracting Party within the limits of the area to which the
         present Convention applies.

Thus, it will be seen that item (iii) has the effect of importing into
the definition, *inter alia*, oil rigs and similar structures and installations.[26]
The '*maritime area*' covered by the Convention is specified in Article
2 as:

(a) Those parts of the Atlantic and Arctic Oceans and the dependent
seas which lie north of 30° north latitude and between 42° west
longitude and 51° east longitude [*Author's Note*: 36° north represents
the southernmost tip of Spain in the Straits west of Gibralter; 42° west
is approximately mid-Atlantic as between the west coast of Ireland
and Newfoundland; and 51° east will include the greater part of the
Barents Sea] but *excluding*:
(i) the *Baltic Sea*[27] and Belts lying to the south east of lines drawn
from Hasenore Head to Gniben Point, from Korshage to Spodsbjerg
and from Gilbjerg Head to Kullen [*Author's Note*: an area of sea
between southern Sweden and off-shore Denmark in the vicinity of
Malmö] and [also excluding]
(ii) the *Mediterranean Sea*[28] and its dependent seas as far as the point
of intersection of the parallel of 36° north latitude (*supra*) and the
meridian of 5° 36′ west longitude [*Author's note*: the southern tip
of Spain, as already indicated] .

The 'maritime area' is also further defined in Article 3 as meaning:
the high seas, the territorial seas of Contracting Parties and waters on
the landward side of the base lines from which the breadth of the territorial
sea is measured, and extending in the case of watercourses, unless otherwise decided[29] by the Commission, created to operate the provisions of
the Convention (*post*), up to the freshwater limit. 'Freshwater limit'
means the place in the watercourse where, at low tide and in a period of
low freshwater flow, there is an appreciable increase in salinity due to
the presence of sea-water.
The Convention establishes a Commission[30] composed of representatives of each of the Contracting Parties, which meets at regular intervals

and also in 'special circumstances', in accordance with its own rules of procedure.

The undertakings by the Contracting Parties as to the agreed method of implementation and application of the Convention are specified in some detail in Articles 4 to 29, and each Party agrees to implement programmes and measures[31] by means of their own State legislation (or, in the case of the Community, by Community legislation), which shall

> include, as appropriate, specific regulations or standards governing the quality of the environment, discharges into the maritime area, such discharges into watercourses as affect the maritime area, and the composition and use of substances and products. *These programmes and measures shall take into account the latest technical developments.* [32]

Thus, it will be seen that the latest techniques are to be seriously considered for application, although it is not mandatory that the 'best technological means' be adopted; and in fact, provision is made for the Contracting Parties to take account[33] of (a) the nature and quantities of the pollutant under consideration; (b) the level of existing pollution; (c) the quality and capacity of the receiving waters of the maritime area; and (d) the need for an integrated 'planning policy' consistent with the requirement of environmental protection. It is as well to interpolate here that the two official texts for the Paris Convention are in French and English. The expression used in the French text is simply *'d'une politique intégrée d'aménagement compatible avec les impératifs de la protection de l'environnement'*, which has a rather wider meaning than *'aménagement du territoire'*, (i.e. town and country planning).

Thus, these provisions, in a sense, introduce many of the discretionary factors (save for that of cost to the polluter) which have traditionally been applied under English and Scots legal provisions.

The Convention also places emphasis on the urgency of the situation, not only by agreeing that the Contracting Parties will place time limits upon the completion of their programmes,[34] but in particular by providing that they will *eliminate*[35] (although, if necessary, by stages) pollution by certain stated substances. The reasons given are (i) because these substances are not readily degradable or rendered harmless by natural processes; *and* (ii) because they may either (a) give rise to dangerous accumulation of harmful material in the food chain, *or* (b) endanger the welfare of living organisms causing undesirable changes in the marine ecosystems, *or* (c) interfere seriously with the harvesting of seafoods or with other legitimate areas of the sea; *and* (iii) because it is considered that pollution by these substances requires urgent action.[36] The substances so specified are:[37] (i) organohalogen compounds[38] and substances which become such compounds when in the marine environment (but excluding 'those which are biologically harmless, or which are rapidly converted in the sea into substances which are biologically harmless'[39]); (ii) mercury and mercury compounds; (iii) cadmium and cadmium compounds; (iv) persistent synthetic materials which may float, remain in suspension or sink,[40] and which may seriously interfere with any legitimate use of the sea; (v) persistent oils and hydrocarbons of petroleum origin.

Although a brief reference will subsequently be made to the European Council Directive of 4 May 1976 on 'pollution caused by certain dangerous substances discharged into the aquatic environment of the Community',[41] it should be observed here that that Directive[42] requires and empowers the European Commission to lay down 'the limit values'[43] which the emission standards[43] must not exceed for the dangerous substances included in the families and groups of substances within List I. List I includes the substances specified in Annex A, Part I of the Paris Convention, *but* it also adds three other groups of substances (although with the general saving clause applicable to the whole of List I families and groups of substances), namely: organophosphorus compounds; organotin compounds; and 'substances in respect of which it has been proved that they possess carcinogenic[44] properties in or via the aquatic environment'.

Article 4 of the Paris Convention, in addition to the undertaking to eliminate pollution by the substances in Part I of the Annex, also provides for the undertaking 'to limit strictly' pollution by substances given in Part II of the Annex, which are distinguished[45] from those in Part I on the basis that 'they seem less noxious or are more readily rendered harmless by natural processes'. For these substances, see footnote 46.

In addition, the parties agree to operate a permanent monitoring system,[47] and the Convention provides that, in addition to the separate responsibilities of each member State, the permanent Commission set up under the Convention will be responsible for continuous monitoring. It has been subsequently agreed that this will be operated through a Joint Monitoring Group to cover both the Paris and the Oslo Conventions. This Joint Monitoring Group meets twice a year, but regrettably, as at early 1979, no permanent joint system has yet been agreed. *The prime difficulty is in arriving at a suitable and agreed system of measurement of pollutants.* The three principal alternative means are (1) by input measurement (rate of discharge into water, etc.); (2) by measurement of concentration of any given pollutant in the affected water; and (3) by measurement of concentration of the given pollutant (or pollutants) in specified marine organisms, such as fish, shellfish, etc. The Paris Convention provides for the use of method (2) above. It should be noted, however, that the three systems at present considered feasible by the European Community are different from the system adopted in the United States of America, where trace elements are taken as indications of unlawful levels of water pollution (see Chapter 8, *ante*). That system, however, would seem to present substantial problems, both in terms of arriving at legal or administrative definitions of 'traces'; and in terms of the degree of strictness of control, bearing in mind that the majority view in the European Community is that the control criterion applied in the United States — namely, that of 'best technological means' — would be unsuitable in terms of economic cost and the 'market philosophy' of the EEC.

The third category of pollutants covered by the Paris Convention is *radioactive substances.* By Article 5, the Contracting Parties undertake 'to adopt measures to forestall and, as appropriate, eliminate pollution of the maritime area from land-based sources by radioactive substances referred to in Part III of Annex A', and to take full account of recommendations, and of the monitoring procedures, initiated by 'appropriate international organisations'. They also undertake to co-ordinate the monitoring and study of radioactive substances.[48] It is not possible here to discuss the general question of the need within Europe for nuclear energy and other uses of radioactive materials, but it should be noted that the Council of Ministers adopted a resolution on energy and the environment as early as 3 March 1975,[49] in which it was provided that the Community and the member States should study the special problems associated with the development of atomic energy, and in particular the danger of radiation and the problems of reactor safety, thermal discharge, radioactive waste and the reprocessing of nuclear fuels. A further Council Directive in June 1976 revised the basic safety standards and made further provision for the management and storage of radioactive waste. The current programme is also concerned with the ecological and health effects of radiation and radioactive particles. The programme is now, of course, of vitally increasing importance, in the light of the fuel crisis resulting from the rise in oil prices.

Finally as to the Paris Convention, it is necessary to make brief reference to certain of the administrative provisions relating to the

permanent Commission set-up under the Convention. This Commission
(which must not of course be confused with the European Commission)
is required:[50] (a) to exercise overall supervision over the implementation
of the Convention; (b) to review generally the condition of the seas
covered by the Convention, the effectiveness of the control measures
and the need for further measures; (c) to fix, if necessary, when proposed
by Contracting Parties bordering on a watercourse, the *position of the
'freshwater limit'*; (d) to draw up measures for the elimination or
reduction of land-based pollution; (e) to make recommendations in
regard to consultations[51] between Contracting Parties in cases of land-
based pollution of types *other than* those listed in Part I of Annex A
(i.e. the substances which are dangerous and/or not readily degradable by
natural processes, etc. *(ante)*; (f) to receive, review and transmit
information[52] as to the permanent monitoring system, assessment of
the effectiveness of any measures taken for reduction of land-based
pollution, and the most detailed information available on the substances
listed in Annex A, Parts I, II and III of the Convention; (g) to make
recommendations as to the amendment of those lists; and (h) 'to discharge
such other functions, as be appropriate, under the terms of the present
Convention'. As to decisions by this Commission, the rules of procedure
are as agreed by the Commission itself, but the Convention provides that,
in regard to programmes and measures for pollution control, and for
research and monitoring, a unanimous vote is required;[53] or, if unanimity
be not achieved, a three-quarters majority vote.[53] In the case of the
adoption of recommendations for Amendments to the lists in Annex A,
a three-quarters majority is sufficient, but these are to be submitted for
the approval of the governments of the Contracting Parties.[54] If a
government is unable to approve such an amendment, this must be
notified to the depositary government (France) within 200 days of the
adoption by the Commission of the recommendation. There is also a
provision[55] that the depositary government shall call a Conference if a
two-thirds majority of the members of the Commission so request, for
the purpose of revising or amending the Paris Convention itself.

   So far as the interpretation or application of the Convention is
concerned, any dispute between Contracting Parties (N.B. including the
EEC) which cannot be settled between them or by conciliation within
the Commission, shall, if requested by both parties, be submitted to
arbitration under the conditions laid down in Annex B of the Convention,
i.e. by reference to an arbitral tribunal. The tribunal is to consist of an
arbitrator appointed by each of the two parties in dispute; and these
two arbitrators will, in turn, appoint the third arbitrator, who shall be

chairman, and shall not be a national of either of the parties to the dispute, nor usually reside in their territory, nor be employed by them, nor have been concerned in the case in any other capacity. The decision of the arbitral tribunal can be by a simple majority.[56] There are provisions for time limits[57] for the appointments, and also to the effect that, if one of the parties fails to appoint an arbitrator within two months of the request by the other party, that other party may inform the *Secretary-General of the United Nations*, who shall designate the chairman of the arbitral tribunal; and upon designation, that chairman will request the party which has not yet appointed an arbitrator to do so within a further two months. If the appointment of that party's arbitrator is still not made within two months, the Secretary-General will then (within a further two months) appoint such an arbitrator. It will thus be seen that a party to a dispute cannot avoid arbitration, but that a refusal voluntarily to participate can result in a total delay of up to six months in the completion of membership of the tribunal.

2.2.2 The Barcelona Convention for the Protection of the Mediterranean Sea against Pollution, 16 February 1976[58]

The Mediterranean Sea and the Baltic Sea were the only two maritime areas of the European Community not included in the Paris Convention. The problems affecting the Mediterranean and the Baltic are substantially different from—and in most cases, more acute than—those affecting the oceans and seas already considered. In the Mediterranean, the absence of any significant tidal movement and oceanic current results in a far more limited natural ability of the sea to degrade pollutants, or otherwise to render them harmless.

The Barcelona Convention having been signed on 16 February 1976, the European Community (as an individual signatory) ratified the Convention on 25 July 1977, and it came into effect in January 1978. However, as at March 1979, only 11 out of the 19 signatories have so far ratified, and difficulties are, no doubt, tending to arise as a result of differences in economic and/or industrial strength between various Mediterranean countries. However, further, very significant steps have been taken by the Contracting Parties at subsequent Conferences initiated by the United Nations Environment Programme at Geneva in 1978 and 1979. These have resulted in a statement of intent that a further Convention specifically relating to land-based marine pollution in the Mediterranean Sea will be drawn up by 1980. The Conferences have been attended by 18 out of the 19 signatories of the Barcelona Convention and the problems recognised relate not merely to industrial waste and silt, but also to very large-scale sewage outfalls from such cities as Athens, where there is virtually no prior treatment. There is also the probability that the flow of water into the Sea from the River Nile will be greatly depleted by inland desert irrigation schemes, and that this will cause further problems in terms of natural degradation of pollutants by the action of the sea itself. It would seem probable that the largest financial contributions towards the extensive engineering and other works and processes for reduction of pollution in the Mediterranean will have to be borne by France, Italy and Spain.

The geographical coverage of the Barcelona Convention is defined[59] as

the Mediterranean proper, including its gulfs and seas, bounded to the west by the meridian passing through Cape Spartel lighthouse at the entrance to the Straits of Gibralter, and to the east by the southern limits of the Straits of the Dardanelles between Mehmetick and Kumkale lighthouses.

The *definition of 'pollution'*[60] differs slightly in wording from that adopted in the Paris Convention,[61] in that, whereas the Paris Convention uses the phrases, *inter alia*, 'harm to living resources and to marine ecosystems, damage to amenities or interference with other legitimate uses of the sea', the Barcelona Convention in this respect refers to 'hindrance to marine activities including fishing, impairment of quality for use of sea water and reduction of amenities'. It would thus appear that there are fine distinctions to be drawn, in that there is, in the latter Convention, no specific prohibition of 'harm . . . to marine ecosystems', although 'living resources' must not be harmed and hindrance of fishing must not occur. However, it is arguable that disturbance of a marine ecosystem will normally tend to affect living resources adversely, including fisheries.

It should also be observed that the Barcelona Convention is concerned not merely with land-based marine pollution, but with the general problems of pollution as affecting the economic, social, health and cultural value of the marine environment of the Mediterranean Sea.[62]

In view of the general approach adopted in this chapter, and of the fact (as already stated) that the Paris Convention is discussed at some length by way of an example of modern thought on suitable international provisions, it is not intended to consider the Barcelona Convention in the same depth. The undertakings by the signatories are broadly similar,[63] and certain selected aspects only will be discussed below.

There is, however, also a specific provision in the Barcelona Convention that nothing in this Convention shall prejudice the codification and development of the Law of the Sea by the United Nations Conference on the Law of the Sea[64] (as and when such material development eventually occurs!).

The Convention provides that the Contracting Parties are to take measures, in conformity with international law, to prevent dumping from ships and aircraft,[65] pollution from ships,[66] or as a result of 'exploration and exploitation of the continental shelf and the seabed and its subsoil'.[67]

With regard to pollution from land-based sources, under the Barcelona Convention it is provided (in Article 8) that:

> The Contracting Parties shall take all appropriate measures to prevent, abate and combat pollution of the Mediterranean Sea area caused by discharges from rivers, coastal establishments or outfalls, or emanating from any other land-based sources within their territories.[68]

The wording used is rather more widely drawn than that in the Paris Convention,[69] which specifies pollution through *watercourses*, or 'from the coast, including introduction through underwater or other *pipelines*, or or from *man-made structures* . . . within the limits of the (maritime) area'. The wording of the Barcelona Convention should, therefore, avoid the controversy which arose subsequent to the Paris Convention as to the intention of the signatories that 'man-made structures' should include oil rigs. It would certainly seem that the Barcelona Convention intends to cover, in addition to rivers (and estuaries), outfalls of all types (not necessarily 'piped' as in the wording of the Paris Convention), all coastal establishments, such as oil refineries, etc. Presumably the undertaking as to avoidance of pollution from oil rigs is also fully covered by the separate reference (under Article 7) to pollution resulting from exploration and exploitation of the continental shelf and the sea bed and its subsoil (substrata).

The Barcelona Convention provides that the Contracting Parties shall '*endeavour*' to establish, in close co-operation with the international bodies which they consider competent, complementary or joint programmes for pollution monitoring, 'and shall endeavour to establish a *pollution monitoring system* for that area'.[70] They also 'undertake as far as possible to develop, and co-operate in, scientific and technological co-operation'.[71] In this sense, the stated intention would appear to be rather less positive and optimistic than that in the Paris Convention.

However, specific Protocols now apply to dumping from ships and aircraft[72] and to co-operation in combating pollution by oil and other harmful substances.

As a general comment, it can be said that the provisions of the Barcelona Convention are less precise in terms of the commitment of each Contracting Party, and that there is reference to future co-operation in seeking to arrive at the formulation and adoption of further Protocols. No Commission or interim Commission has been set up; and secretariat functions are to be carried out by a designated 'organization'.[73] This organisation is, in fact, the United Nations Environment Programme[74] which, in addition to convening meetings of the Contracting Parties, transmitting information and considering enquiries, is to perform any functions assigned to it by Protocols to the Convention, or as may be assigned by the Contracting Parties. The same organisation is also responsible to ensure the necessary co-ordination with international bodies. Thus, the 'organization', unlike a Commission, has no powers of commitment on behalf of the Contracting Parties. Additional Protocols,[75] may, however, be adopted by the Contracting Parties at a diplomatic conference convened at the request of two-thirds of the 19 signatories; but the voting requirement for an amendment to the Convention itself is a three-quarters majority.[76] Disputes as to the interpretation or practical application of the Convention, or of the Protocols under the Convention, may be submitted to *arbitration* as laid down in Annex A to the Convention.[77] These provisions are in all essentials similar to those for arbitration under the Paris Convention (*ante*).

Any of the Contracting Parties may withdraw from the Convention on expiry of 90 days from notification of withdrawal being received by the Depository[78] (the Spanish Government). In the case of the Paris Convention, however, withdrawal can be after expiry of one year from notification (to the French Government), but not earlier than 7 May 1980.

2.2.3 The Helsinki Convention on the Protection of the Baltic Sea Against Pollution, of 22 March 1974[79]

The European Community is not, as yet, a Contracting Party to this Convention, although the Federal Republic of Germany and Denmark are individual signatories, and a Council of Ministers Decision of 21 June 1977[79] authorised the European Commission to open negotiations for Community accession. However, as a result of political opposition, the Eastern Bloc signatories have so far refused to open negotiations. Since the legislation adopted within the European Community has been selected as the main example, in order to limit the length of this chapter, the terms of the Helsinki Convention will not be discussed.

2.3 Other European Community Provisions as to Water Pollution, Having a Bearing upon Pollution of the Marine Environment

It is clear that the provisions of the Paris Convention and—perhaps to a lesser extent—the Barcelona Convention, combined with their related Council Decisions, are theoretically fully capable of achieving the reduction, or elimination, of marine pollution from land-based sources. However, in particular due to the fact that 85 or 90 per cent of such pollution emanates from freshwater sources, legal control over these fresh waters within the land mass of the European Community is also essential. These controls are applied through various Directives and Regulations (*vide* distinction between these, *ante*). The principal Directives are as to: the Dumping of Wastes at Sea (discussed elsewhere in this book); the quality of surface water intended for abstraction of drinking-water in the member States;[80] the quality of bathing water;[81] and pollution caused by certain dangerous substances discharged into the aquatic environment of the Community.[82] The principles applied in these Directives will be outlined later.

In addition to these general Directives, the Council also approved two directives in 1973 relating to detergents[83] which are based upon standards set for minimum biodegradability (i.e. the extent to which the chemical constituents can be broken down by living organisms). These standards are higher than any previously set by an international organisation, but it is significant that a 'gentleman's agreement' exists to permit the United Kingdom to apply lower levels—no doubt party due to the shorter rivers and more rapid run-off. There is also a Council Decision of 27 September 1977[84] in regard to a concerted three-year project for the treatment and use of sewage sludge, as to sludge stabilisation and odour, sludge dewatering, and environmental problems connected with sludge use.

A further Directive was issued on 18 July 1978[85] on the quality of fresh waters needing protection or improvement in order to support fish life, which relates to salmonidae (i.e. salmon, grayling, trout) and cyprinids (carp, bream, chub, minnows, etc.) as well as pike, perch and eels. This Directive requires member States to *designate* the waters to which it will apply, and to fix limit values[86] for certain specified parameters applicable to designated waters, such as temperature, dissolved oxygen, amonium, phosphorous, nitrites, chlorine, zinc and copper within certain guide values given in the Directive.

Lastly, as already mentioned, the Community is a Contracting Party to the Convention on the Protection of the Rhine against Chemical Pollution.

However, full implementation of these Directives cannot be achieved until the practical mechanics of control (including application of sanctions against polluters) can be finally designed and laid down. The Resolution of the Council on the Continuation of the Environmental Action Programme, passed on 17 May 1977,[87] provides a helpful summary of the situation in regard to control of pollution generally. Chapter 2 of the Resolution covers 'Prevention and reduction of pollution of fresh and sea water'. The priorities include: preparation of a Directive on the protection of ground water; the setting of 'quality objectives' for water (originally intended as early as 1973); the definition of methods of measurement and frequency of sampling; the definition of methods for attaining and maintaining present and future water quality; investigation and definition of the minimum long-term quality requirements; exchange of information between monitoring networks; the establishment of 'emission limit values', maximum deadlines and quality objectives for the initial group of dangerous substances given in List 1 of Directive 76/464/EEC (i.e. as to pollution caused by dangerous substances into the aquatic environment of the Community[88]); together with establishment of a surveillance and monitoring procedure. The setting up of water treatment plants, and of pollution and monitoring stations by the Community (in addition to member States) was envisaged for completion within a 15-year period (i.e. up to 1992).

The 'market philosophy' of the European Community requires that — with certain exceptions such as, notably, the Common Agricultural Policy (C.A.P.) — there should be free market competition within the Community. This principle tends to militate against the creation of false trading positions by means of subsidies, and thus it is not acceptable to seek to achieve pollution control by offering grants or loans for the purpose of enabling outdated production or processing plants either to go out of business or to adopt modern technological means of pollution prevention. Thus, in general, industrialists will be required to find their own technical and financial means of meeting pollution control requirements at such time as they are brought into force. There are, however, certain exceptions approved, or likely to be approved, by the Council — as, for example, in the long-term policy of moving pulp mills at present situated in inland forest areas down to river estuaries in order to prevent the present high-density (but non-toxic) pollution caused by wood pulp in certain rivers.

The requirement that industrialists will eventually have to meet the cost of pollution control whilst seeking to remain competitive is, not unnaturally, producing much extended discussion by member State governments and the *Union des Industries de la Communauté Européene* (UNICE), which represents the national confederations of industrialists of each member State. The present view (early 1979) of UNICE appears to be that the European Commission is tending to be over-persuaded by the Commission's scientific advisers and research bodies, in terms of requiring too exacting a standard, both as to the levels set for water quality and the complicated and delicate analytical techniques which industrialists may be required to apply without suitable laboratory facilities. Before considering the practicalities which stem from this, it is necessary to refer briefly to the aims of each of the relevant Directives.

2.3.1 The Convention on the Protection of the Rhine against Chemical Pollution

was signed in Bonn on 3 December 1976, and concluded by a Council Decision of 25 July 1977.[89] The provisions are almost identical to those in the Council Directive (not Decision) of 5 May 1976 as to pollution caused by certain dangerous substances discharged into the Aquatic Environment of the Community (*post*).

2.3.2 The Directive Concerning the Quality of Surface Water Intended for the Abstraction of Drinking-Water in the Member States[90]

This Directive results from a proposal of the Commission and opinions of the European Parliament and of the Economic and Social Committee. The use of water resources for human consumption is increasing, and it is necessary to protect public health; but it is also the case that any disparity between the quality required of surface water intended for the abstraction of drinking-water may create unequal conditions of competition (i.e. be contrary to the 'market philosophy'). Accordingly, the Directive provides for the definition of parametric values for water intended for this purpose. The Directive provides that the parametric values will have to be assessed by 'the methods currently being worked out for water sampling and for measuring the parameters defining physical, chemical and microbiological characteristics', which are to be adopted 'as soon as possible'.

In view of the fact that the comparatively high standards set for drinking-water (*per se*) even prior to the Directive will be such as to have no material effect upon the sea if such water flows into it (and, in any event, from the microbiological point of view, would be materially

sterilised by salt water), it is not proposed to discuss this particular
Directive in any detail. It may, therefore, suffice to say that the Directive
is concerned only with[91] 'surface water' which is fresh water used or
intended for use in the abstraction of drinking-water. The directive
specifically excludes ground water, brackish water, and water intended
to replenish water-bearing beds. It does, however, specifically apply to
drinking-water within supply networks (i.e. also after any supposed
purification process).[91]

The Directive divides surface water according to 'limiting values' into
categories A1, A2, and A3,[92] which correspond to the standard methods
of *treatment* (i.e. in increasing intensity of treatment) given in Annex I
of the Directive. These three categories of treatment also correspond to
the three different qualities of surface water, as to physical, chemical
and microbiological characteristics set out in a table of parameters in
Annex II. The table shows 46 items. The first six provide parameters for
pH (acidity), colouration, suspended solids, temperature, conductivity
and odour, but (at present) only colouration (after simple filtration)
and temperature have mandatory, as opposed to 'guide', parameters.
Items 7 to 30 include quantities in milligrammes per litre of eighteen
elements and five salts; but mandatory parameters are so far set only
for fluoride, dissolved iron, copper, zinc, arsenic, cadmium, total
chromium, lead, selenium, mercury, barium and cyanide, together with
nitrates and sulphates. There is provision that, in the case of water in
shallow lakes not exceeding 20 metres in depth, or in virtually stagnant
surface water, the parameters as to nitrates, dissolved iron, manganese
and phosphates can be waived by member States,[93] provided that the
decision is forthwith notified to the Commission.

Items 31 to 40 of the table relate to various chemical compounds, of
which mandatory parameters apply to phenols, dissolved and emulsified
hydrocarbons (after extraction of petroleum ether), and total pesticides
(parathion, BHC and dieldrin).[94] The final items, numbers 41 to 46,
relate to living or dead organisms or waste from such organisms. Mandatory parameters apply to faecal coliforms, faecal streptococci, and to
salmonella (except that in the case of salmonella no mandatory
parameter applies to category A3 water).

2.3.3 The Directive Concerning the Quality of Bathing Water[95]

relates to *all* water for bathing with the exception of water intended for
therapeutic purposes or used in swimming pools.[96] '*Bathing water*' is
further defined as

all running or still fresh waters or parts thereof and sea water in
which:
— bathing is explicitly authorised by the competent authorities of
   each Member State, or
— bathing is not prohibited and is traditionally practised by a large
   number of bathers.[97]

It will be seen that the member States, by this Directive, are responsible
for the quality of waters—fresh water or sea-water—in which any substantial number of bathers are in the habit of bathing. Thus, it is not
open to a member State to limit their responsibility by specifying areas
deemed to be the most suitable: they must accept traditional practice
and cater for it. Indeed, the Directive states that 'bathing area' means
any place where bathing water is found.[98] There is, however, a limitation,
in the sense that the samples of water required to be taken by the
member State need only commence two weeks before the annual 'bathing season', as defined in relation to the periods when 'large numbers of
bathers can be expected', etc.[99]

The Directive—as in the case of the Directive relating to drinking-
water (*ante*)—specifies, in the Annex, certain (although more limited)
parameters which are to be applied in regard to both microbiological
quality and physio-chemical quality; and the frequency of sampling
and method of analysis and inspection are laid down.

The general conditions affecting 'bathing water' will clearly vary
according to whether it is fresh running water; fresh still water; or
whether it is coastal water cleansed by tidal or ocean current flows; or
areas of sea water in which self-cleansing is much below average, as in
the Mediterranean or the Baltic Sea. It would be fair to say that, both
prior to adoption of the Directive and subsequently, in terms of
implementation, the requirements as to bathing water have been deemed
to be of far greater importance to the coastal States of the Mediterranean
having tourist resorts, than to the colder, northern States, particularly in
areas where there may be extensive industry or large populations and a
scattered distribution of 'traditional' bathing areas. Equally, the existence
of traditional bathing by a large number of bathers in fresh water may
have no learly logical distribution. It exists for example, in the Serpentine
Lake in London (where the 'bathing season' traditionally includes
Christmas Day and New Year's Day!); but may also exist in rivers subject
to industrial and sewage contamination. In this sense, no doubt the
stretch of the River Thames at Oxford, known as 'Parson's Pleasure', may
be favoured in view of the improvement in the waters of the Thames over

the past two decades. However, 'responsible authorities' (regional Water Authorities) may well meet with severe difficulties where the traditional attraction to large numbers of bathers has stemmed from the surrounding scenery, rather than from quality of the river or sea-water itself.

The member States are required to take all the necessary steps specified in Articles 3 to 7, and the Annex, of the Directive, before December 1985. However, the Directive may be waived[100] in the case of certain parameters indicated in the Annex table in the case of exceptional weather or geographical conditions; or where 'natural enrichment' causes a deviation from the values (parameters) prescribed in the Annex. This natural enrichment is defined[100] as the process whereby, without human intervention, a given body of water receives from the soil certain substances contained therein. But, where a member State waives these provisions of the Directive, they are required immediately to notify the Commission (i.e. the European Commission, since this does not relate to an international Convention having a separate commission). However, the Directive provides for the setting up of a Committee on Adaptation to Technical Progress[101] consisting of representatives of the member States, but 'chaired by a representative of the Commission'. This representative is to present to the Committee drafts of the successive measures to be adopted for achieving such technical progress, and the Committee shall deliver its opinion on the drafts within time limits set by the chairman, according to the degree of urgency. Opinions must be adopted by a majority of 41 votes, the chairman not voting, these votes of the member States being weighted according as provided[102] in Art. 148(2) of the Treaty of Rome.[103]

The Commission must then adopt the opinion of the Committee and propose the measures included in it to the Council of Ministers, which shall act by a qualified majority.[103] It will be evident from the procedure laid down that the Committee on Adaptation to Technical Progress will tend to become the forum for the expression of differing views as to the controls to be applied, and also as a body in which negotiation may also take place. As in the case of other Directives relating to water pollution, the industrial sector (and in this case, potentially also the municipal sector in regard to sewage) tends to oppose the degree of control and the exactitude required in terms of sampling, inspection and analysis.

2.3.4 The Directive on Certain Dangerous Substances Discharged into the Aquatic Environment

is the last of the Community's Directives which should be referred to here.[104] This applies to inland surface water, territorial waters, internal

coastal waters and to ground water.[105] As to the latter, the statement
of intent indicates that, 'subject to certain exceptions and modifications,
this Directive shall be applied to discharges into ground water pending
the adoption of specific Community rules in the matter.' Such rules are
currently under consideration. (It should, however, be observed that
statutory controls as to ground water already exist in the United Kingdom and some other member States.)

'Inland surface water', for the purposes of this directive, means 'all
static or flowing fresh surface water situated in the territory of one or
more Member States'.[106] 'Internal coastal waters' means 'waters on the
landward side of the baseline from which the breadth of territorial waters
is measured, extending, in the case of watercourses, up to the fresh-water
limit'.[107]

'Discharge' means[108] the introduction into the waters referred to in
paragraph 1 (of Article 1) of any substances in List I or List II of the
Annexe, *except* discharges of dredgings, operational discharges of ships
in territorial waters and dumping from ships in territorial waters. The
latter two operations are covered under separate Directives (*q.v.*). The
definition of 'pollution'[109] is essentially similar to that given in the
Paris Convention (*ante*).

The dangerous substances are divided into those in List I (the 'black-
list'), discharge of which is to be eliminated (although if necessary, by
stages); and those in List II (the 'grey list'), discharge of which must be
very strictly limited and, in some cases, eventually totally eliminated.
The List I substances are selected due to toxicity, and for persistence
and/or bioaccumulation. They are organohagonic, organophoephoric
and organotin compounds; those with carcinogenic properties; mercury
and cadmium and their compounds; persistent mineral oils and hydrocarbons of petroleum origin; and persistent substances which may
float, remain in suspension or sink and which may thus interfere with
any use of the waters.

The List II substances include certain metals and their compounds—
such as zinc, copper, nickel, chromium, lead, selenium, arsenic, antimony,
molybdenum, titanium, tin, barium, berylium, boron, uranium, vanadium,
cobalt, thalium, tellurium, and silver, together with those biocides not
included in List I; substances that have a deleterious effect on the taste
or odour of those products for human consumption which are derived
from the aquatic environment; toxic or persistent organic silicon compounds; inorganic phosphorous compounds and elemental phosphorous;
non-persistent mineral oils; cyanides and fluorides; and substances
such as ammonia and nitrates which have an adverse effect on the

oxygen balance.

With regard to List II, it should be noted that provision is also made for control of substances which are liable to 'give rise to such substances (as listed) in water'; but, equally, that allowance is also made so as to permit those toxic organic compounds of silicon which are either 'biologically harmless', or which are 'rapidly converted in water into harmless substances'. Even so, the Directive states that the provisions relating to List II represent only a first step towards complete elimination of the substances listed.[110]

The basis of control is that all discharges of substances in List I (or which are 'liable to contain' such substances) require prior authorisation by the competent authority of the member State concerned;[111] and such authorisation must lay down emission standards either into the waters (as defined in Article 1) direct, or into sewers.[112] In the case of discharging sources which already exist, the authorisation must state a time limit within which the discharge must be discontinued.[113] And, in any event, authorisations may be granted only for a limited period, although they may be renewed,[114] taking into account any changes in the limit values as to concentration of the toxic substance in any discharge, and 'maximum quantity of such substance expressed as a unit of weight of the pollutant per unit of the characteristic element of the polluting activity (e.g. unit of weight per unit of raw material or per product unit)'.[115] These limit values are laid down by the Council.

As to discharges into ground water, the substances in List I must not be emitted; and the substances in List II may be emitted subject to the provisions as already stated for List III; *but* the limits on both lists do not apply to domestic effluents or to discharges injected into deep, saline or unusable strata.[116] It is also provided that these provisions shall cease upon the implementation of the intended separate Directive on Ground Water.

In regard to the substances in List I, the Council of Ministers is to lay down deadlines for compliance;[117] and as to those in List II, the member States are required to establish programmes in which emission standards are specified in order to achieve the quality objectives for water 'in accordance with Council Directives, where they exist'; that is, such directives as, for example, those concerning the quality of surface water intended for the abstraction of drinking-water (*ante*), or the quality of bathing water (*ante*).[118] However, it must be stressed that the programmes 'shall take into account the latest *economically feasible* [author's italics] technical developments.'[119]

Finally, in terms of practical application of controls, the competent authority in each member State is required to draw up an inventory of all discharges of substances covered by List I. This information is to be available to the Commission, and the member States are charged with providing the Commission, at its request, with details of authorisations, the results of the inventory, the results of monitoring by the national monitoring network, and additional information as to the national programmes put into effect.[120] It is, however, specifically provided that the Commission and the competent authorities of member States, their officials and other servants, shall not disclose information 'covered by an obligation of professional secrecy'[121] (trade secrets). Flexibility is also provided for within the Directive as to future alterations to the standards required and to transfer of substances from List I to List II.

2.4 Further Observations as to the Practical and Trade Effects of Existing and Intended Directives in Regard to Water Pollution

In legislating for control of water pollution, the Community has in mind both that the 'market philosophy' requires that the means of prevention shall not be achieved by grants or subsidies (although with certain very limited exceptions); and second, that the 'polluter pays' principle shall apply. However, as to the latter, further complications arise.

It is accepted by industry that the cost of prevention (or of changes in manufacture so as to avoid or minimise pollutant waste) shall be met by industry itself; but the process of achieving this in practice presents problems. At present there is a difference of views in the Community as to whether there should be control by prohibition (absolute or partial) enforced by sanctions such as fines or imprisonment; or whether the eventual aims can be achieved by means of 'pollution charges' – or taxes, which will be payable by every identified polluter, and may be at an annually increasing rate. The purpose of this type of pollution charge is essentially to *coerce* the industrialist into improving his technology: it is not essentially to provide public funds for expenditure by a public authority, since this would tend to reduce the liability placed upon the polluter to rectify the situation. However, although this method of control is being visualised and developed in regard to surface water, it does not appear that it is being applied to pollution of sea-water, brackish water, or, potentially, to ground water.

As already mentioned, water pollution can consist either of toxic pollutants (of various degrees of toxicity), or of substances which are not in themselves toxic, but which pollute by sheer volume and/or permanence. The requirements as to methods of legal control are essentially different for these two categories. It may be permissible for member States to apply their own level of charge or sanction within their territory to high volume pollution; but the sanctions applicable to toxic pollution ought to be fixed throughout the European Community.

In particular in regard to water pollution, the Commission is seeking to find a basis on which to harmonise the systems of pollution taxes which already exist in France, the Federal Republic of Germany, Italy and the Netherlands (the latter being based on raising funds for public expenditure on anti-pollution measures), and to encourage the other member States to apply pollution charges. In this respect, the United Kingdom significantly relies upon the cost to an industrialist of rectification as a sufficient indirect tax. However, there is a Communication from the Commission to the Council annexed to the Council Recommendation of 3 March 1975 regarding cost allocation and action by public authorities on environmental matters.[122] The standards are as already mentioned (*ante*), and reference is made to the need to avoid distortion of trade competition and to the acceptance of the 'polluter pays' principle. However, it is also stated in paragraphs 5 to 7 of the Annex:

5. Depending on the instruments used and without prejudice to any compensation due under national law or international law, and/or regulations to be drawn up within the Community, polluters will be obliged to bear:
(a)   expenditure on pollution control measures (investment in anti-pollution installations and equipment, introduction of new processes, cost of running anti-pollution installations, etc.), even when these go beyond the standards laid down by the public authorities;
(b)   the charges.
The costs to be borne by the polluter (under the 'polluter pays' principle) should include all the expenditure necessary to achieve an environmental quality objective, including the administrative costs directly linked to the implementation of anti-pollution measures. The cost to the public authorities of constructing, buying and operating pollution monitoring and supervision installations may, however, be borne by those authorities.[123]
6. Exceptions to the 'polluter pays' principle may be justified in limited cases:
(a)   Where the immediate application of very stringent standards or the imposition of substantial charges is likely to lead to serious

economic disturbances, the rapid incorporation of pollution control costs into production costs may give rise to greater social costs. It may then prove necessary:

— to allow some polluters time to adapt their products or production processes to the new standards;

— and/or to grant aid for a limited period and possibly of a degressive nature.

Such measures may in any case, apply only to existing production plants[i] and existing products.

(b)   Where, in the context of other policies (e.g. regional, industrial, social, and agricultural policies or scientific research and development policy), investment affecting environmental protection benefit from aid intended to solve certain industrial, agricultural or regional structural problems.

Aids referred to under (a) and (b) may, of course, only be granted by Member States in compliance with the provisions on State aid set out in the Treaties establishing the European Communities, and in particular Articles 92 *et seq.* of the EEC Treaty. In applying Articles 92 *et seq.* of the EEC Treaty to these aids, account will be taken of the requirements which such aids satisfy as regards environmental protection.

(i) The enlargement of the transfer of existing production plants will be considered as the creation of new plants where this represents an increase in production capacity.

7. The following shall not be considered contrary to the 'polluter pays' principle[ii]:

(a)   financial contributions which might be granted to local authorities for the construction and operation of public installations for the protection of the environment, the cost of which could not be wholly covered in the short term from the charges paid by polluters using them. In so far as other effluent as well as household waste is treated in these installations, the service thus rendered to undertakings should be charged to them on the basis of the actual cost of the treatment concerned;

(b)   financing designed to compensate for the particularly heavy cost which some polluters would be obliged to meet in order to achieve an exceptional degree of environmental cleanliness;

(c)   contributions granted to foster activities concerning research and development with a view to implementing techniques, manufacturing processes and products causing less pollution.

(ii) This list may be modified by the Council on a proposal from the Commission.

Nevertheless, the question of whether to use the United States of America's principle of rendering all sources of pollution unlawful which fail to meet the test of 'best practicable *technology*' or to apply – as currently in the European Community – the test of 'best practicable *means*', which thus includes the element of reasonable cost, is still the subject of debate. However, it should be observed that, in England, where best practicable means has been applied as a defence in nuisance for many years, the effect has more recently tended to be greatly cut down by the magistrates' courts[124] in relation to statutory nuisances under the Public Health Acts. The tendency has been to apply increasingly strict criteria as to what in fact constitutes a 'nuisance' (for example the degree of frequency of a smell) and, at the same time, to say that where there is a known technological solution, this must be applied if an order to cease production is to be avoided. This type of decision is, of course, to be distinguished from the decisions reached by the Alkali and Clean Air Inspectors (under the Control of Pollution Act 1974 and the Health and Safety at Work, etc., Act, 1974), where an element of discretion applies in some cases.

In practical terms, there is the difficulty of ensuring adequate monitoring of waters as part of the 'policing' necessary for effective pollution control. There are contrasting approaches as between those member States which prefer to legislate for inflexible rules and heavy penalties, but informally (on occasion) to turn a blind eye; whereas other member States tend to favour the rules being made applicable at the discretion of the responsible authority, but with firm application of penalties when these are transgressed.

It is also probable that the Community will decide against the use of 'best practicable technology' as a defence, because this would tend to 'freeze' technological development and experimentation, in that industrialists would continue with outdated but safe methods, rather than risk new, and possibly improved, methods of manufacture. In addition, it would seem that the Commission could usefully examine the question of a means of measurement of cost-effectiveness in terms of economic cost relative to environmental cost, which could reasonably be expected to be applied by manufacturers when developing technological prevention of pollution.

With regard to the long-term plans for relocation of certain types of industry which cause widespread pollution (such as pulp mills situated in inland forest areas which are to be moved down river estuaries so as to avoid high-volume — but non-toxic — pollution), although assistance will be given through regional subsidies (*ante*), it should be stressed that this assistance will be of the short-term, or 'once only' type, in order to comply with the 'market philosophy' which requires that industry should not, in general, be subsidised as to production costs.

At the same time, it is probable that a system of *progressive* pollution charges will be adopted (i.e. increasing year by year), since they appear to have had the effect of inducing technological progress in France and Germany. However, at the same time, the Netherlands applies a deliberate system of spreading the cost of pollution by levying charges not merely upon industrialists in a given locality, but also upon local inhabitants, and applying the funds thus raised to prevention or reduction of the pollution, or to relevant technological research.

In terms of the application of pollution charges, the use of stated *norms* as to specified pollutants found in waters subject to control — whereby the charges apply whenever a norm is exceeded — there is the practical advantage that no distinction is drawn between a polluter who (even accidentally) exceeds the norm by a minute fraction, and one who very substantially exceeds it. It may be that scales of norms and related charges could be developed, and this would have the effect of providing a positive inducement to decrease any pollution above the given norm.

It has been observed by industry that the parameters so far applied to industry have increased production costs in the sectors affected by as much as 40 per cent.

In parallel with the systems of pollution charges, however, it should be borne in mind that the administrative systems developed for pollution control should be designed so as to *avoid delay* in industrial development; and that any unnecessary limitations upon industrial waste which result from unusual local conditions (such as tidal flow or oceanic currents) should be avoided, if it is clear that no harmful effects can result. A knowledge of this aspect of local geophysical and climatic conditions will emerge as the ecological mapping study (*ante*) is pursued.

The setting of limits or norms and the requirements as to sampling and laboratory testing by the scientists employed by the Commission have, in the general view of industry, tended to be too exacting, and often impractical under industrial and industrial laboratory conditions. Industry therefore hopes that this scientific approach may be appropriately adjusted. The development of further means of environmental protection should be an evolutionary and practical process, in which it is sought to achieve the best results in return for a reasonable outlay, and in which intended developments in industrial processes are visualised and reasonably provided for.

There are also problems to be resolved in terms of the responsibilities of regional Water Authorities within member States. The principle adopted is that each Water Authority should be economically autonomous, and that they are responsible for the quality and availability of water supplies within their individual regions. This requires exercise of controls over drainage (including land drainage and natural run-off into rivers), together with a liability for development of purification plants and supply systems. However, in the United Kingdom and some other member States, the present national system provides for transfer of water from one regional Water Authority to another in times of localised or national water shortages. Problems therefore arise in regard to charging for water supplied to another Water Authority, and as to capital costs of any works necessary to provide this flexibility of supply. Further thought as to this will be required within the Community.

3 CONCLUSIONS

A final comment may be ventured. In considering legislation for the reduction of land-based marine pollution, the Council of Ministers, and the Commission (together with the European Parliament, although in a debating and non-legislative capacity), are faced with the problem of reconciling quite widely divergent interests between individual member States. Clearly it is of great importance to inland States that they retain reasonable means of disposal of waste material; whereas, in the case of coastal States, a substantial proportion of their national economies and local populations may be dependent upon the sea, including coastal waters. In addition, some coastal States are primarily dependent upon tourism, whereas others have substantial industrial areas on the coastline. Thus, even although the coastal States of the Community are already tending to acknowledge a degree of common interest, the extent of control of water pollution actually desired by each member State will vary.

The divergence of certain interests between member States has, of course, been provided for by the existence of, on the one hand, Council Decisions and Regulations made by the Commission (which have both immediate and binding effect); and, on the other hand, the use of Council Directives, which state merely the principles which are to be incorporated in legislation by each member State within a specified time.

As a result of this system, however, it is evident from the investigation made prior to writing this chapter that it is at present by no means easy to establish with firm precision what legal constraints apply to any given industrial site or locality within the European Community. Thus, for example, an industrial investor who is contemplating location of a new plant in the Community will find it essential to exercise the utmost care in seeking legal and scientific advice as to *each and every* of the alternative sites he may have in mind.

There is also the problem that the precise chemical and other parameters to be applied under several Directives and Decisions are not yet laid down, and that the exact level and form of control will be of considerable importance in making financial forecasts for alternative sites, more particularly bearing in mind that the preparatory process may well tend to take in excess of three years even before actual site works are commenced. By way of example, it could be disastrous for an industrial investor if, after a suitable site had been selected and financial forward planning and architectural design work completed, a watercourse upon which the site relied became designated as waters for protection or improvement in order to support (freshwater) fish life.[85]

It is, therefore, of importance that—at least for the next, say, five years—improved sources of completely up-to-date information be made available, since this will consist of both (i) a more accurate economic assessment in the course of forward planning of European industry; and (ii) a related, more precise forecast of the environmental effects of such industrial proposals, both generally and in terms of land-based marine pollution.

*It is important to realise that this approach to local planning, and to water pollution control, includes the intention of eventual legislation requiring that the principle of Environmental Impact Assessment (EIA) should apply to those development projects—whether by private developers or by public authorities—which are likely to have a significant effect upon the environment. In this respect the following eight paragraphs have been revised at the latest possible stage in publication. At this time, the Environmental and Consumer Protection Service of the European Commission is continuing a further initiative in order to provide a draft Directive of a type likely to be acceptable to the Commissioners collectively, and to the Council of Ministers, subject to consultation with the European Parliament and the Environmental and Social Committee of the Community.

Although the writer has had the opportunity of studying with some care the latest two provisional drafts, it would be inappropriate to seek to state their content in detail. The following general observations may, however, be of some value. The Directive may possibly come into effect by the end

of 1980, or by mid-1981, and it is drafted to allow a further two years
before compulsory introduction within all Member States. In this
respect it is worthy of note that, although the United Kingdom has
appeared to adopt an attitude of, to some extent, disapproving disinterest,
France has a fully established statutory system of EIA, and West Germany
is equally active.

The approach adopted by the current provisional draft Directive (the
19th) is to provide that EIA will apply initially to development projects,
but will later extend to land-use plans (even including forestry areas or
the re-allocation of agricultural holdings which continue in agriculture),
and to development plans. It is noted in the Directive (and rightly) that
there is considerable disparity between measures of planning and control
between Member States, which produces unfavourable competitive conditions in terms of the Common Market; and the draft therefore aims to
provide a comprehensive process for assessment of environmental effects,
whilst at the same time leaving Member State governments a degree of
latitude as to the more precise methods and parameters to be applied.

The provisions may be subject to amendment during the legislative
process, but the following observations are selected as of relevance to
water pollution. EIA may be applied not merely to construction projects,
but (in specified types of production, etc., etc.) also to the operation, or
modification (e.g. extension, change of use, or decommissioning), of
installations and facilities. It is, however, the Member State governments
who are likely to decide which development projects will tend, due to
their nature, size and location, to have significant effects upon the
environment. And it is also worthy of note that, in the case of trans-
frontier effects, although the relevant adjoining Member States are to
consult, the actual EIA report need not be provided by one Member
State to the other. This latitude probably results from the need to
achieve enactment of the Directive without increased opposition within
the Council of Ministers and the other bodies concerned.

Provision is also made to enable Member States to modify the detail
required in an EIA in relation to projects of certain types specified in the
Directive, but subject to prior reference to the Commission.

In general terms, the draft Directive sets out in some detail the means
of assessment of the local environment and the effect—or alternative
effects—thereon of the proposed development and feasible alternatives,
in accordance with criteria for assessment as specified in an Annex to
the Directive; and it also aims to impose a duty upon the competent
authority (e.g. Local Planning Authority, Water Authority, etc.) to make
public the details of the EIA, either separately or as part of the decision

relating to planning permission, and to enable the general public to express their views. The manner in which they may formally offer these views is not, however, laid down.

The Articles in the draft Directive include requirements that planning permission for projects likely to have significant effects on the environment shall not be granted without a prior environmental assessment, and that the developer shall be responsible for submitting an Environmental Impact Statement at the time of making his planning application. This statement is to describe the project and also the local environment likely to be affected, and it will assess the 'important' environmental effects the project is expected to have. In addition, the developer will be required to justify the reason for choosing the particular development alternatives, and to provide a non-technical summary of these items. However, it should be remarked that there is provision for the protection of trade secrets.

The draft Directive also requires that the competent authorities ensure adequate protection of the environment in relation specifically to water, air, climate, soil, flora and fauna (ecosystems); the built environment; the use of raw materials; and the present and potential use of these elements as resources.

The draft Directive then provides, in Annexes, a very full list of types of industry and of construction projects which will be subject to assessment. Finally, screening criteria are specified, and the content of Environmental Impact Statements listed. The extent of detail required may be illustrated by the need to forecast, by type and quality, the expected residual liquid, solid and gaseous pollutants; radiation; noise; vibration; and odours. It should, however, also be noted that provision is now made for a not strictly environmental item to be included, namely the envisaged effects upon employment and income in the locality.

Notes

1. 'Qualified majority votes', weightings: Belgium 5; Denmark 3; Germany 10; France 10; Ireland 3; Italy 10; Luxemburg 2; Netherlands 5; United Kingdom 10.
2. Berne, 26 April 1963.
3. European Documentation Periodical 1977/6, The European Community's Environmental Policy.
4. O.J. No. C 112, 20.12.1973, p. 1, supplemented by Agreement of 15 July 1974, O.J. No. C 86, 20.7.1974, p. 2.
5. O.J. No. C 139, 13.6.1977, p. 6, Title 1, Restatement of the Objectives and Principles of a Community Environment Policy.
6. O.J. No. C 139, 13.6.1977, p. 19. June 1978, Title III, Chapter 1, section 1.
7. June 1978, Title III, Chapter 1, Section 1, para. 90, The Ecological Mapping Project.
8. O.J. No. C 139, 13.6.1977, p. 6, Title 1, para. 13.
9. O.J. No. C 139, 13.6.1977, p. 7, Title 1, para. 17.
10. O.J. No. L 194, 25.7.1975, p. 1.
11. Council Decision 76/311 of 15 March 1976; O.J. No. L 74, 20.3.1976, p. 36.

12. Council Decision 76/161 of 8 December 1975; O.J. No. L 31, 5.2.1976, p. 8.

13. European Documentation Periodical 1977/6, The European Community Environmental Policy, p. 12.

14. Draft proposals only, Commission of the European Communities, March 1978.

15. 15 February 1972, the Convention for the Prevention of Marine Pollution by Dumping from Ships and Aircrafts.

16. 13 November 1972, the Convention on the Dumping of Wastes at Sea.

17. May 1974.

18. June 1974.

19. February 1976.

20. Eutrophication = an over-accumulation of living organisms, such as algae, etc.

21. 1973 Action Programme, Part II, Title I, Chapter 6, Section 1.

22. But Art. 24 provides for accession of additional Contracting Parties, and Art. 25 for withdrawal after 2 years from 6 May 1978, upon giving 1 year's notice. It is also provided that the European Community is entitled to the number of votes equal to its Member States parties to the Convention (8); but that these individual States may exercise their voting rights independently on any issue.

23. Council Decision 75/437/EEC of 3 March 1975; O.J. No. L 194, 25.7.1975, p. 5 (Convention given in Annex to Decision); and Council Decision 75/439/EEC of 3 March 1975; O.J. No. L 194, 25.7.1975, p. 22. These Decisions also relate to the previously related Community policy in Part II, Title 1, Chapter 6, Section 1, of the First Environmental Action Programme (1973); and also to the Community's participation in the Interim Commission, and (later) Commission, set up under Resolution III of the Paris Convention.

24. The countries which have not ratified as at March 1979 are: Belgium Germany, Iceland, Ireland, Luxemburg and Spain.

25. O.J. No. L 194, 25.7.1975, p. 5. Annex: Paris Convention, Article 1.

26. Although there has been subsequent controversy over the intended meaning as between the Contracting Parties.

27. On 14 March 1977, the European Commission forwarded to the Council a recommendation for a Decision concerning the opening of negotiations with a view to the accession of the Community to the Convention of 22 March 1974 on the protection of the environment of the Baltic Sea area, but the Eastern Bloc States who were signatories to the Convention have so far not agreed to accession of the European Community as such.

28. See the Barcelona Convention (*post*).

29. Art. 16(c) of the Paris Convention.

30. Art. 15.

31. Art. 4.

32. Author's italics. Article 4, para. 3.

33. Art. 6, paras. 2(a) to (d).

34. Art. 4, para. 3.

35. Art. 4, para. 1(a).

36. Art. 4, para. 1 and Annex A to the Convention, Part I.

37. Annex A, Part I, numbered paras. 1 to 5.

38. Organohalogen = e.g.: compounds of chlorine, fluorine, etc.

39. Annex A. Part I, para. 1.

40. This phrase is presumed to be explanatory, unless it is intended to provide for the exclusion of substances which 'take off' from the water surface!

41. O.J. No. L 129, 18.5.1975, p. 23.

42. In Art. 6 of that Directive and Annex List I thereto.

43. See later for discussion of the Directive of 4 May 1976.

44. Cancer-producing.
45. Annex A, Part II, first paragraph.
46. Substances listed in Part II of Annex A:

    1. Organic compounds of phosphorus, silicon, and tin and substances which may form such compounds in the marine environment, excluding those which are biologically harmless, or which are rapidly converted in the sea into substances which are biologically harmless.
    2. Elemental phosphorus.
    3. Non-persistent oils and hydrocarbons of petroleum origin.
    4. The following elements and their compounds:
    Arsenic
    Chromium
    Copper
    Lead
    Nickel
    Zinc.
    5. Substances which have been agreed by the Commission as having a deleterious effect on the taste and/or smell of products derived from the marine environment for human consumption.

47. Arts. 11 & 17 of Paris Convention.
48. Arts. 5 & 11, and Part III of Annex A.
49. O.J. No. C 168, 25.7.1975, p. 2.
50. Art. 16(a) to (h) of Paris Convention.
51. Consultations provided for, Art. 9 of Paris Convention.
52. Under Arts. 11, 12 & 17.
53. Art. 18, para. 3.
54. Art. 18, para. 4.
55. Art. 27.
56. Under these arbitration provisions there is no Referee, since the decision is by simple majority.
57. Art. 4 of Annex B.
58. O.J. No. L 240, 19.9.1977, p. 1.
59. Art. 1 of Barcelona Convention.
60. Art. 2(a) of Barcelona Convention.
61. Art. 1 of Paris Convention.
62. Preamble to the Barcelona Convention.
63. Art. 4 of Barcelona Convention.
64. Art. 3, para. 2 of Barcelona Convention, and Resolution 2750 C (XXV) of the General Assembly of the United Nations.
65. Art. 5, Barcelona Convention.
66. Art. 6, Barcelona Convention.
67. Art. 7, Barcelona Convention.
68. Art. 8, Barcelona Convention.
69. Art. 3(c) of Paris Convention.
70. Art. 10 of Barcelona Convention.
71. Art. 11.
72. Arts. 24 & 26, and Protocol, O.J. No. L 240, 19.9.1977, p. 12, and substances listed in Annexes I, II and III, to the Protocol.
73. Art. 2(b) of Barcelona Convention.
74. Art. 13.
75. Art. 15.
76. Art. 16, para. 3.
77. O.J. No. L 240, 19.9.1977, p. 10.

78. Art. 28 of Barcelona Convention.
79. Not reported in Official Journal.
80. Council Directive 75/440/EEC of 16 June 1975, O.J. No. L 194, 25.7.1975, p.26.
81. Council Directive 75/160/EEC of 8 December 1975, O.J. No. L 31, 5.2.1976, p.1.
82. Council Directive 76/464/EEC of 4 May 1976, O.J. No. L 129, 18.5.1976, p. 23.
83. Council Directives 73/404/EEC and 73/405/EEC; O.J. No. L 347, 17.12.1973, pp. 51 and 53.
84. Council Decision No. 77/651 of 22 September 1977; O.J. No. L 267, 19.10.1977, p. 35.
85. Council Directive No. 78/659 of 18 July 1978; O.J. No. L 222, 14.8.1978, p. 1.
86. As listed in Annex I to the Directive.
87.
88.
89. Council Decision No. 77/586 of 25 July 1977; O.J. No. L 240, 19.9.1977, p. 35.
90. Council Directive 75/440/EEC of 16 June 1975; O.J. No. L 194, 25.7.1975, p. 26.
91. Art. 1, para. 1.
92. Art. 2.
93. Art. 8(d).
94. These are, *by agreement with manufacturers*, not produced for general sale in the United Kingdom.
95. Council Directive 76/160/EEC of 8 December 1975; O.J. No. L 31, 5.2.1976, p. 1.
96. Art. 1, para. 1 of Directive.
97. Art. 1, para. 2(a).
98. Art. 1, para. 2(b).
99. Art. 1, para. 2(c).
100. Art. 8.
101. Art. 10.
102. Art. 11.
103. Art. 11, para. 3. For weighting as to voting rights see footnote 1.
104. Council Directive 76/464/EEC of 4 May 1976; O.J. No. L 129, 18.5.1976, p. 23.
105. Art. 1.
106. Art. 1, para. 2(a).
107. Art. 1, para. 2(b).
108. Art. 1, para. 2(d).
109. Art. 1, para. 2(e).
110. Art. 2.
111. Art. 3, para. 1.
112. Art. 3, para. 2.
113. Art. 3, para. 3.
114. Art. 3, para. 4.
115. Art. 6, para. 1.
116. Art. 4.
117. Art. 6, para. 4.
118. Art. 7, para. 3.
119. Art. 7, para. 4.
120. Arts. 11, 12 & 13.
121. Art. 13, para. 4.
122. 75/436/Euratom, ECSC, EEC; O.J. No. L 194, 25.7.1975, pp. 182 *et seq.*
123. However, the Confederation of British Industry takes the view that industry should be assured of having the opportunity of undertaking these measures at a cost calculated by industry, provided that such measures would be effective.
124. Magistrates' courts – the lowest level of criminal or quasi-criminal courts (in England, as distinct from Scotland), presided over either by a stipendiary magistrate or, more usually for the purposes of the Public Health Acts, by three lay magistrates.

11 LIABILITY FOR OIL POLLUTION: UNITED STATES LAW

Allan I. Mendelsohn and Eugene R. Fidell

1. INTRODUCTION

As the United States approaches the last two decades of the twentieth century, few areas of its law prove to be as complex as that governing liability for oil pollution. The subject involves not only some of the most fundamental principles of tort law but also admiralty doctrines largely judge-made, legislative enactments and international conventions.[1] If further complexity were needed, the interplay of federal and State legal systems amply provides it. Because of the varied sources of law, and because the matter of oil pollution has still not been the subject of comprehensive federal legislation, it is perhaps a larger task than ought to be the case to render a brief account of American liability law in this area. None the less, the present chapter will provide at least a cursory overview of the major elements of federal legislation, subject to the *caveat* that critical elements of the story—particularly those having to do with State regulation—have been woven in with perhaps less emphasis than they merit.

2 THE LIMITATION OF LIABILITY ACT OF 1851

The first statute we shall consider is not directly addressed to oil pollution as such; rather it deals with the general liability of shipowners, and in this way has an impact on liability for pollution.

The basic maritime liability law of the US today is a federal statute enacted in 1851 and amended since that time only so as to bring about some improvement in recoveries available by or on behalf of passengers who suffer personal injury or death in a ship disaster.[2] So far as concerns vessel-owner liability for oil pollution damage, therefore, the principal governing law in the United States today is, with some exceptions, one that was adopted almost 130 years ago.[3] It is particularly fitting that the present chapter begin with a consideration of this statute, as it was largely inspired by English legislation,[4] even if its judicial construction was soon 'liberated' from its English antecedents.[5]

S. 183(a) of the Shipowner's Limitation of Liability Act provides that, in the event of an accident 'occasioned or incurred without the privity or knowledge' of the vessel-owner and involving damage to property, the owner's liability shall not 'exceed the amount or value of the interest of such owner in such vessel, and her freight then pending'. In 1871, the Supreme Court held that 'value' as used in s. 183(a) meant the vessel's value after, not before, the accident, thus severely restricting the potential recovery where, for example, the vessel for which limitation is sought lies unsalvageable on the ocean bottom.[6] A series of Supreme Court decisions in 1885 specifically held that even though an owner, after an accident, is fully compensated by insurance, the funds so received are not a part of the vessel's value and hence need not be made available to pay claims.[7] In the case of the *Torrey Canyon* disaster off the coast of Britain in March 1967, a United States court, following the letter of s. 183(a), would have been able to award damages of no more than $50 – the value of the single lifeboat that was salvaged from the disaster.[8]

Given the potentially harsh results from this statute as it has been construed, it is not surprising that ways have been tried – and found – to 'break the limits'. One of these ways is simply by showing that the damage was done or incurred with 'the privity or knowledge of' the owner – in which event the owner's liability becomes unlimited. No one can really divine the meaning of the words 'privity and knowledge' in s. 183 and, indeed, Professors Gilmore and Black, in their excellent treatise on *The Law of Admiralty*, have called the terms 'devoid of meaning . . . empty containers into which the courts are free to pour whatever content they will'. 'Like an accordion', Gilmore and Black added, the Act 'can be stretched out or narrowed at will' depending on the facts of the particular case and the philosophy of the judge hearing the case.[9] Another method of breaking the limits is for the court to ignore the traditional admiralty rule holding that the law of the *forum* governs limitation[10] and, instead, to invoke the law of the vessel's flag, assuming, of course, that this approach permits a higher recovery than would otherwise be available under s. 183.[11] But any court viewing its function as judicial rather than legislative ultimately faces a philosophical problem in continuing the search for ways to ignore the limitation. This is particularly true at present, because the Congress has had ample opportunity to amend, update, and/or repeal s. 183, but has failed to do so in any meaningful manner except indirectly and in the limited context of water pollution control legislation.[12] It is the development of the law in this area to which we next turn.

3 FEDERAL WATER POLLUTION CONTROL ACT AMENDMENTS OF 1972

Amending and expanding upon various provisions of earlier Congressional enactments,[13] the Congress in 1972 adopted comprehensive legislation pertaining to virtually all aspects of water pollution and Federal Government recovery for costs of clean-up.[14] The Federal Water Pollution Control Act Amendments of 1972, recently codified and updated in 33 U.S.C. ss. 1251-1376 (now known as the *Clean Water Act*), declares it to be a national goal that the discharge of pollutants into the navigable waters of the US be eliminated by 1985.[15] S. 1321(b)(1) declares it to be the policy of the US that there be

> no discharges of oil or hazardous substances into or upon the navigable waters of the United States, adjoining shorelines, or into or upon the waters of the contiguous zone or in connection with activities under the Outer Continental Shelf Lands Act or the Deepwater Port Act of 1974, or which may affect natural resources belonging to, appertaining to, or under the exclusive management authority of the United States (including resources under the Fishery Conservation and Management Act of 1976).

Oil is defined as 'oil of any kind or in any form, including, but not limited to, petroleum, fuel oil, sludge, oil refuse, and oil mixed with wastes other than dredged spoil'. A discharge is defined as including, but not limited to, 'any spilling, leaking, pumping, pouring, emitting, emptying or dumping'.[16]

A vessel-owner, operator or any other person in charge of a vessel, including a demise charterer, is subject to a $5,000 civil penalty for any unlawful discharge, and the Environmental Protection Agency (EPA) may, under Amendements passed in 1978, commence a civil action for penalties up to $250,000. In the event of failure by such person immediately to notify designated agencies of the United States Federal Government of a discharge, the statute prescribes a maximum fine of $10,000 or imprisonment for not more than one year, or both.[17]

When a discharge occurs, the US Government is authorised to undertake and complete the clean-up unless the President determines that the vessel-owner or operator can and will properly perform the task.[18] In any instance where the Federal Government cleans up, the vessel-owner or operator is strictly liable to the Government for its 'actual costs incurred' in the clean-up—but not to exceed, in the case of all vessels (other than inland oil barges), the greater of $150 per gross ton or

$250,000. A tanker of 200,000 tons could thus be subject under this provision to a potential clean-up liability to the Federal Government of $30 million.[19] A somewhat lower limit of the greater of $125 per gross ton or $125,000 is provided for inland oil barges.[20] To meet this potential liability, the statute requires vessel- and barge-owners to show evidence of financial responsibility (by insurance, surety bonds, self-insurance or other means) in sums sufficient to cover the limits.[21] This latter programme is administered by the Federal Maritime Commission.

Vessel- or barge-owners may avoid liability entirely by showing that the discharge was

> caused solely by (A) an act of God, (B) an act of War, (C) negligence on the part of the US Government, or (D) an act or omission of a third party without regard to whether any such act or omission was or was not negligent, or any combination of the foregoing clauses.[22]

An 'act of God' is defined as 'an act occasioned by an unanticipated grave natural disaster'.[23]

On the other hand, the right to limitation can be extinguished and full liability assessed against the vessel-owner in any case where the Federal Government can show that the discharge 'was the result of willful negligence or willful misconduct within the privity and knowledge of the owner'.[24] The US is also expressly authorised to bring an action for the recovery of its costs 'in any court of competent jurisdiction', which does not necessarily preclude bringing actions in foreign courts should it ever be necessary to do so, and should the *forum* State have no domestic doctrine barring access to its courts in these circumstances.[25] Finally, s. 1321(k) of the statute carries over the $35 million revolving fund which was originally created by s. 11(k) of the 1970 Water Quality Improvement Act. This is a fund financed by Congressional appropriation and available to assist the Government in carrying out its obligations under the Act, including the payment of expenses incurred on those occasions when the Government cleans up but, for one reason or another, cannot recover its costs, or when a private party cleans up and then, in accordance with s. 1321(i), sues the Government to recover its clean-up costs.[26]

It should be noted that the foregoing provisions establish vessel-owner liability only for claims by the Federal Government for its costs of clean-up.[27] None of them cover claims of private citizens whose property is damaged as a result of an oil spill. These types of claims can always be asserted under State law, though to do so necessarily raises issues concerning, for example, the applicability of the limits of the 1851 Limitation of Liability Act, and the varying damages compensable under different state statutes. On the other hand, s. 1321(o)(2) of the 1972 Act, adopting almost the verbatim text of s. 11(o)(2) of the 1970 Water Quality Improvement Act, declares that '[n]othing in this section shall be construed as preempting any State or political subdivision thereof from imposing any requirement or liability with respect to the discharge of oil or hazardous substance into any waters within such State'.

In the landmark case of *Askew* v. *American Waterways Operators, Inc.*, 411 U.S. 325 (1973), a unanimous Supreme Court held that the federal legislation did not pre-empt or preclude the State of Florida from enacting its own Oil Spill Prevention and Pollution Control Act imposing strict liability on vessel-owners and on operators of on-shore terminals for any damages incurred by the State or by private persons as a result of an oil spill in the State's territorial waters. The Court reasoned that, as the federal legislation governed only claims by the Federal Government for recovery of its clean-up costs, it would not be proper

> to allow federal admiralty jurisdiction to swallow most of the police power of the States over oil spillage—an insidious form of pollution of vast concern to every coastal city or port and to all the estuaries on which the life of the ocean and the lives of the coastal people are greatly dependent.[28]

The Court expressly declined to reach the question whether damages under the Florida Act would be subject to the 1851 Limitation of Liability Act.[29]

4 THE TRANS-ALASKA PIPELINE AUTHORIZATION ACT OF 1973

Specifically allowing private parties to recover for oil pollution damage without regard to the 1851 Limitation of Liability Act, the Trans-Alaska Pipeline Authorization Act (the TAPS Act) was adopted by the US Congress in November 1973.[30] As for vessel-owners, the Act extends only to 'oil that has been transported through the trans-Alaska pipeline' and on vessels operating 'between the terminal facilities of the pipeline and ports under the jurisdiction of the United States'. However, the Act holds the vessel-owner 'strictly liable without regard to fault . . . for all damages, including clean up costs, sustained by any person or entity, public or private, including residents of Canada, as a result of discharges of oil' from vessels.[31]

In addition to vessel-owner liability, the TAPS Act also imposes liability on the holder of the pipeline right-of-way for any 'damages in connection with or resulting from activities along or in the vicinity of [such] right-of-way'. Like the vessel-owner, the holder of the right-of-way is 'strictly liable to all damaged parties, public or private, without regard to fault for such damages'.[32] As for defences to liability, the TAPS Act treats the vessel-owner identically to the holder of the right-of-way, limiting both their defences only to those cases where damages 'were caused by an act of war or by negligence of the US or other governmental agency'.[33]

The TAPS Act therefore significantly expands the ambit of protection beyond the clean-up costs available only to the Federal Government under the Federal Water Pollution Control Act Amendments of 1972. In addition, it eliminates two of the defences generally available to vessel-owners under the 1972 Act (discharges caused either by 'an act of God' or by 'an act or omission of a third party without regard to whether any such act or omission was or was not negligent'). In this latter regard, moreover, the TAPS Act expressly recognises those circumstances where damage is caused by the conduct or activity of a third party. Unlike the 1972 Amendments, however, the TAPS Act provides that such circumstances may not be used by the vessel-owner or the holder of the right-of-way as a defence against all claims, but only against claims by that third party, and then only if negligence by that third party can also be proven.[34]

The liability of the right-of-way holder is limited to $50 million for any one incident.[35] If the damage is caused by a vessel, however, this Act for the first time incorporates the concept of a liability-type fund to assure both that all damaged parties can be compensated for damages they might suffer and to shift and apportion the economic responsibility for such damages to those who benefit from the carriage of oil by vessels, namely the shipowner, the oil industry and the oil consumer.[36] Financed by a 5-cent per-barrel fee levied by the pipeline operator upon the oil owner at the time the oil is loaded on a vessel, the Trans-Alaska Pipeline Liability Fund will ultimately total $100 million to be used for the purpose of providing compensation in circumstances where proven damages exceed $14 million. Up to $14 million, liability is borne by the vessel-owner or operator 'jointly or severally'. Beyond that amount, the Fund takes over the responsibility of compensation up to its $100 million ceiling.[37] By August 1978, the Fund contained $13 million.[38]

Like the 1972 Water Pollution Control Act Amendments, the TAPS Act expressly addresses the subject of pre-emption of State laws, and similarly declares that its liability provisions 'shall not be interpreted to preempt the field of strict liability or to preclude any State from imposing additional requirements'.[39] Finally, and again following the precedent—indeed, referring to the precise provision—of the 1972 Amendments, the TAPS Act requires vessel-owners to show evidence of financial responsibility (by insurance, surety bonds, self-insurance or other means) sufficient to cover their potential liability of $14 million.[40]

5 THE INTERVENTION ON THE HIGH SEAS ACT OF 1974

In November 1969, the Brussels International Legal Conference on Marine Pollution adopted the International Convention Relating to Intervention on the High Seas in Cases of Oil Pollution Casualties. On 20 May 1970, President Nixon transmitted this Convention to the Senate for its advice and consent.[41] In 1971, the Senate gave its advice and consent, and in 1974, the Congress enacted the Intervention on the High Seas Act for the purpose of implementing and incorporating into US statutory law the rights, duties and responsibilities of the US under the Convention.[42] Recently codified in 33 U.S.C. Sections 1471-87, the Act authorises the Secretary of the Department of Transportation (in which the Coast Guard ordinarily operates) to take such measures on the high seas as may be necessary to prevent, mitigate, or eliminate 'grave and imminent danger to the coastline or related interests of the United States from pollution or threat of pollution of the sea by oil'.[43] Among the listed 'interests' of the US are 'fish, shellfish, and other living marine resources, wildlife, coastal zone and estuarine activities, and public and private shorelines and beaches'.[44]

When the Secretary determines that a 'grave and imminent' danger exists, he is authorised to 'remove, and, if necessary, destroy the ship and cargo which is the source of the danger'.[45] However, actions taken by the Secretary must be reasonable given the circumstances, and the Act makes the US liable, by way of a suit in the Court of Claims, for damages caused by actions beyond 'those reasonably necessary' in any particular situation.[46] Finally, the Act makes available to the Secretary, for the purpose of financing 'actions and activities' undertaken by the Government pursuant to the Act, the resources of the revolving fund of $35 million created by the 1970 Water Quality Improvement Act and carried over by the 1972 Federal Water Pollution Control Act Amendments.[47]

6 THE DEEPWATER PORT ACT OF 1974

Because the waters surrounding most of the United States (and especially adjacent to Gulf and East Coast ports) are not sufficient to accommodate supertankers requiring depths of up to 100 feet, Congress in late 1974 enacted the Deepwater Port Act, establishing a licensing and regulatory programme governing off-shore deep-water port development beyond the territorial limits and off the coasts of the US.[48] Codified in 33 U.S.C. ss. 1501-24, the Act prohibits oil discharges from any of the facilities of the port or from any vessel operating to or from the port or in a surrounding area technically denominated a 'safety zone'. Designated by the Secretary of Commerce 'subject to recognized principles of international law', this 'safety zone' is an area 'of appropriate size around and including any deepwater port for the purpose of navigational safety'.[49]

A discharge in violation of the Act either by a vessel-owner or operator or by a licensee of the port is punishable by a civil penalty not to exceed $10,000 for each violation. Failure to notify the Government of a discharge is punishable by a fine of not more than $10,000 or imprisonment for not more than one year or both.[50]

As for the nature of liability, the Deepwater Port Act adopts the same strict liability with the same limited defences as were adopted the year before in the TAPS Act. Both the vessel-owner and the licensee are strictly liable 'without regard to fault' for damages, which are broadly defined as 'clean up costs and . . . damages that result from a discharge of oil'. The only defences available are, as in the TAPS Act, 'an act of war, or negligence on the part of the Federal Government in establishing and maintaining aids to navigation'. In addition, if it can be shown that the damages resulted solely from the negligence of the party asserting a claim for the damage, neither the owner nor the licensee would be liable to that party.[51] The vessel-owner's liability arises from any 'discharge of oil from such vessel within any safety zone, or from a vessel which has received oil from another vessel at a deepwater port'.[52] The licensee's liability arises when the discharge is 'from a vessel moored at such deepwater port' or from the deep-water port itself.[53]

As for the limits of liability, the owner and the licensee are subject to unlimited liability if it can be shown that the discharge occurred because of 'gross negligence or willful misconduct within [his] privity and knowledge'. Absent such a showing, the limitation is $50 million for the licensee and the lesser of $150 per ton or $20 million for the vessel owner.[54]

For damages in excess of those compensated by the vessel-owner or the licensee, the Act again follows the precent of the TAPS Act by creating a $100 million fund, called the Deepwater Port Liability Fund. Financed by a 2 cent per barrel fee collected by the licensee 'from the owner of any oil loaded or unloaded at the deepwater port', the Fund, like the owner and licensee, is also liable 'without regard to fault'.[55] But unlike the owner and the licensee, the Fund's liability appears to be absolute in the sense that, though the Act allows the Fund a defence against claims by a party whose negligence caused the damage, it does not allow the Fund to invoke either of the other two defences, namely act of war and negligence of the Federal Government in establishing and maintaining aids to navigation. It thus appears that, at least with regard to this latter point, the Deepwater Port Act represents an expansion of the scope of liability from earlier federal legislation in the oil pollution area. Compare, for example, the provision governing the scope of the TAPS Fund liability in 43 U.S.C. s. 1653(c)(2), exonerating that Fund from liability where the damage was caused by an act of war or government negligence.

Another innovation adopted in the Deepwater Port Act which appears to have no precedent in earlier federal pollution legislation is a provision which: (i) allows the Attorney General of the United States to institute a class action 'on behalf of any group of damaged citizens he determines would be more adequately represented as a class in recovery of claims', and (ii) authorises the Secretary of Transportation, acting 'on behalf of the public as trustee of the natural resources of the marine environment', to sue and recover for damages to such resources and then to apply any sums recovered 'to the restoration and rehabilitation of such natural resources'.[56]

Expansive though these provisions are, the Deepwater Port Act, like the TAPS Act, has only very limited applicability. It was at least partially for this reason—though also to attempt to arrive at some element of uniformity in the face of the continuing profusion of varying federal and State funds as well as varying federal and State laws on the subject—that attempts began, following the Deepwater Port Act, to enact as federal legislation what has since become colloquially known as the 'superfund' bill.[57] The purpose of such a measure is to establish a 'superfund' together with a comprehensive legal regime governing liability and compensation, public as well as private, for all damages and clean-up costs caused by oil pollution, no matter the source and no matter the location.[58] But the enactment of such legislation is fraught with difficulties—not only because of its inherently controversial nature, but also because, as more laws on the subject continue to be enacted with increasingly stringent terms, the pressure on any new comprehensive law to equal or exceed the most protective of the terms and levels of all the other laws becomes virtually irresistible. And it is in this context that, before examining the options under consideration for the comprehensive law, we must first examine in some detail the most recent law which the Congress has at this writing enacted in the oil pollution context.

7 THE OUTER CONTINENTAL SHELF LANDS ACT AMENDMENTS OF 1978

Enacted in the closing days of the 95th Congress, the Outer Continental Lands Act Amendments of 1978 (the 1978 OCS Act) amend the 1953 Outer Continental Shelf Lands Act[59] by establishing a new statutory regime for the management of oil and natural gas resources of the outer continental shelf and by adopting provisions to expedite the systematic development of the shelf but at the same time protect the marine and coastal environment.[60] It should be kept in mind that, while as a land mass the OCS is almost one-third the size of the United States, only approximately 3 per cent of the US continental margin, or some 14.4 million acres, have so far been leased for oil and gas development.[61]

Nevertheless, in passing the 1978 Act, Congress seemingly anticipated a much more active programme of exploration and exploitation and, to meet the risks of such a programme, Congress devised and adopted an unprecedentedly broad system to govern pollution damages, clean-up and liability. Applicable to owners and operators of any off-shore facility in the outer continental shelf or vessels operating over the shelf and carrying OCS oil, which 'causes or poses an imminent threat of oil pollution', the Act sets up a system of strict liability under which, except for costs of removal, a vessel-owner is potentially liable for damages up to a limit of $300 per ton (subject to an overall ceiling), and the off-shore facility owner is potentially liable up to a limit of $35 million.[62]

As far as removal costs are concerned—and these are broadly defined to include 'clean-up costs' and any other costs 'incurred' under subsection (c), (d) or (1) of s. 311 of the Federal Water Pollution Control Act[63] and section 5 of the Intervention on the High Seas Act[64]—s. 304(d) of the OCS Act expressly precludes the applicability of all 'limitations, exceptions, or defenses'. It thus renders the vessel or facility owner (as the case may be) absolutely liable, and without limitation, for 'all costs of removal

incurred by the Federal Government or any State or local official or
agency in connection with a discharge of oil from any offshore facility
or vessel'.[65] This may well be the first instance of unlimited and
absolute liability in a federal statute.

Somewhat like the TAPS and the Deepwater Port Acts, though more
capable of being relied upon by damaged parties (not only because of
its unique use of the word 'primarily', but also because it includes the
violation of federal standards and regulations as a trigger for extinguish-
ing the right to limitation), s. 304(b) of the OCS Act provides that the
owner or operator of the vessel or the facility may lose the right to
limitation, and thus also be liable without limit for damages other than
removal costs, when

> the incident is caused *primarily* by willful misconduct or gross
> negligence, within the privity or knowledge of the owner or operator,
> or is caused *primarily* by *a violation*, within the privity or knowledge
> of the owner or operator, *of applicable safety, construction, or
> operating standards or regulations* of the Federal Government
> [emphasis added].[66]

On the other hand, and except for removal costs, a vessel or facility
owner may avoid liability entirely if it can be shown that the incident
was

> caused solely by an act of war, hostilities, civil war, or insurrection,
> or by an unanticipated grave natural disaster or other natural
> phenomenon of an exceptional, inevitable, and irresistible character
> ... [or if it was caused] solely by the negligent or intentional act
> of the damaged party or any third party (including any government
> entity).[67]

Other than the defences enumerated in the 1970 Water Quality Improvement Act, as they were carried over into the 1972 Water Pollution
Control Act Amendments,[68] no subsequent federal statute dealing with
oil pollution damage allows acts of God (for example, a natural disaster)
and negligence of third parties to operate as complete defences to
liability—though the required proof that the incident be caused 'solely'
by one of these factors reduces the potential sweep of the defence.[69]

Damage recoveries, referred to as 'claims for economic loss', are
specifically permitted not only for removal or clean-up costs of the
federal and local governments, but also for such elements of damage as

(A)   injury to, or destruction of, real or personal property;

(B)   loss of use of real or personal property;

(C)   injury to, or destruction of, natural resources;

(D)   loss of use of natural resources;

(E)   loss of profits or impairment of earning capacity due to injury to, or destruction of, real or personal property or natural resources [so long as the claimant 'derives at least 25% of his earnings from activities which utilize the property or natural resource'] ; and

(F)   loss of tax revenue ['by the Federal Government and any State or political subdivision thereof'] for a period of one year due to injury to real or personal property.[70]

In addition, there is another unique provision in the Act rendering the vessel and facility owner liable 'without regard to the limitation of liability [of $300 per ton or $35 million respectively] . . . for interest on the amount paid in satisfaction of the claim for the period from the date upon which the claim is presented to such person to the date upon which the claim is paid'.[71]

Finally, where the damages might exceed the limits or where defences might exonerate the vessel or facility owner from liability, the 1978 Act creates an Offshore Oil Pollution Compensation Fund not to exceed $200 million, to be financed by a fee of not more than 3 cents per barrel imposed on the owner of oil when such oil is produced on the OCS.[72] The $200 million is not an inflexible ceiling, as the Act author- ises the Government to borrow '[i] f at any time the moneys available in the Fund are insufficient to meet the obligations of the Fund'.[73] The Act also expressly provides that 'the Fund shall be liable, without limit- ation, for all losses for which a claim may be asserted under [the Act] . . . to the extent that such losses are not otherwise compensated'.[74]

The Act additionally requires the promulgation of regulations requiring vessel and facility owners to show financial responsibility 'to satisfy [their] maximum amount of liability' subject to the limitations.[75] The Act also provides civil and criminal penalties for failure to give notice to the Government of a discharge or for failure to comply with the financial responsibility requirements.[76] Other sections of the Act include detailed provisions governing such matters as suits by the Attorney General, class actions by private citizens suffering damages, claims settlement procedures, rights of the Fund to sue and be sued, subrogation, jurisdiction and pre-emption. Without analysing these additional sections, it should suffice simply to say that this Act

contains by far the most comprehensive and far-reaching provisions that
have appeared to date in any federal legislation dealing with oil pollution.
And if only because it was the most recent federal enactment, it is
probably the most significant backdrop for Congress's current consideration of the various legislative proposals for a comprehensive 'superfund'
act.

8 PROPOSED NEW FEDERAL LEGISLATION

In a sense, perhaps the most important provision of the Deepwater Port
Act, discussed above, was s. 18(a), which called for a detailed study to
be performed by the Attorney General, in consultation with other
agencies, regarding 'methods and procedures for implementing a uniform
law providing liability for clean up costs and damages from oil spills
from Outer Continental Shelf operations, deepwater ports, vessels, and
other ocean-related sources'. This provision was eloquent evidence of
Congressional dissatisfaction with the piecemeal character of its past
efforts; and the Report that was eventually submitted by Attorney
General Levi proved to be influential in shaping subsequent proposals
for legislation.[77]

Following on the heels of the Justice Department's Report, a host
of superfund Bills was introduced by various Senators and Representatives during the 94th and 95th Congresses. Three emerged with some
possibility of enactment at the close of the 95th Congress. These were
H.R. 6803, which was passed by the House; the version of S. 2083
which was reported out by the Senate Committee on Commerce, Science
and Transportation (hereafter Senate Commerce Committee); and the
version of S. 2083 which was reported out by the Senate Committee on
Environment and Public Works (hereafter Senate Environment Committee). But the press of business at the close of the 95th Congress was
such that, apart from the Outer Continental Shelf Lands Act Amendments, no comprehensive oil pollution legislation was adopted. In
addition, the thorny substantive differences that continued to emerge
during the final days of the Congress among the supporters of the
various bills made it unlikely that, absent a willingness for substantial
compromise on all sides, any new oil pollution legislation could be
enacted even during the relative calm of a new Congress.[78]

Because of the comprehensive nature of the Bills, plus the fact that
none was ultimately enacted, little purpose would be served by a detailed
analysis of their respective provisions. But so that the reader may gain
some understanding of the overall directions in which US pollution law
is apparently moving and of the controversies surrounding that movement,

we shall very briefly outline the various areas of general agreement and disagreement common to the Bills.[79]

8.1 Vessel-Owner Liability

There seems to be general agreement, at least in the Senate, that the limit of liability for oil tankers will be the same $300 per gross ton (with no ceiling) as was enacted in the 1978 OCS Act. A tanker of 200,000 dwt would thus be exposed to a potential liability of $60 million—exactly double the $30 million potential liability to which that tanker is exposed under the current Clean Water Act.[80] Regarding vessels other than tankers, the Senate Committee on Environment prefers the same $300 per gross ton, while the Senate Commerce Committee prefers $150 per gross ton—subject, in each case, to no ceiling. For its part, and so far as oil tankers are concerned, the House of Representatives is likely to accept the $300 per gross ton limit, while professing to subject it to a $30 million ceiling.

8.2 Liability of Facilities

Both Senate Committees as well as the House are reportedly prepared to adopt the limit of $50 million presently in the Clean Water Act for on-shore and off-shore facilities and for deep-water ports.[81] There is also apparent general agreement that such a limit may be reduced in the event a determination is reached that a lower limit would be more appropriate for a particular class or category of facility. In this latter respect, however, the Senate Environment Committee prefers a 'floor' or minimum liability exposure of at least $8 million.

8.3 The Fund and its Funding

Both Senate Committees and the House are in agreement that there should be an Oil Spill Fund to compensate for oil pollution damage and that such Fund, like the OCS Fund, should be in the amount of $200 million, financed by a fee of up to 3 cents per barrel to be imposed on each barrel of crude oil received at any refinery or terminal for import to or export from the US. There is similar agreement that, once established, this Fund will supersede all the Funds established under other federal laws to provide oil pollution liability protection. These include the TAPS Act Fund, the Deepwater Port Fund, the OCS Fund, the revolving fund established under the 1970 Water Quality Improvement Act and carried over in the 1972 Water Pollution Control Act Amendments, and the availability of that latter fund under the Intervention on the High Seas Act.

As in the OCS Fund, the $200 million amount would not be limited, because there is agreement in principle that the Fund may borrow money if the damages exceed $200 million or any other amount that may be in the Fund at the time an incident occurs. The $200 million amount, which the entire Congress now seems inclined to accept, is so much higher than the $39 million fund provided for in the 1971 International Convention on the Establishment of an International Fund for Compensation for Oil Pollution Damage (or even the increase to $78 million permitted by the Convention if three-quarters of the states parties thereto so agree) that it is now highly unlikely that the US will ratify that Convention.[82]

8.4 Hazardous Substances

The Senate Environment Committee, drawing on the 1972 Water Pollution Control Act Amendments and its inclusion of 'hazardous substances' within the protective ambit of the clean-up and liability provisions, prefers similarly to include 'hazardous substances' in any new law that may now be adopted. The Environment Committee also wishes to set up a second fund, separate from the Oil Spill Fund, to cover clean-up costs and damage compensation resulting from spills or discharges of hazardous substances. The Environment Committee has not yet suggested a monetary level nor a fee system to finance this second fund, but the Bill calls for an Administration study (to be completed in 18 months) which would make recommendations on these points. The Senate Commerce Committee, like the House, limits its Bill and proposed Fund to pollution caused by oil.

8.5 Terms, Scope and Nature of Liability

The Environment Committee prefers to employ s. 311 (now codified in 33 U.S.C. s. 1321) of the 1972 Water Pollution Control Act Amendments, as amended by the 1977 Clean Water Act, as the basis for setting out the terms, scope and nature of any system of liability adopted in new pollution legislation. The Environment Committee suggests that, as s. 311 has been part of US law since its adoption as s. 11 of the 1970 Water Quality Improvement Act, there is a persuasive reason not to fashion a new system. The Senate Commerce Committee and the House, for their part, prefer to adopt a new system. Space does not permit a detailed comparison of the proposed systems, but it appears that there is general agreement that, at least as pertains to oil pollution damage, strict liability should be continued. The differences seem to emerge principally in the phraseology of the rights that are entitled to protection as well

as the defences and exceptions that may be invoked to escape liability.
As between the two Senate Bills, neither is significantly more pro-
industry or pro-environment than the other. As between those two
Bills and the House Bill, however, Senator Muskie and the Committee
on Environment are understood to believe the differences to be substantial and significant enough to warrant serious public attention and
discussion.[83]

8.6 Pre-emption

The extent to which State law should be permitted to operate in the
area of oil pollution liability has long been one of the most troublesome
issues between the industry, the environmentalists and the US Government. It is even more contentious today than it was in 1970, when
Congress, in adopting s. 11(o)(2) of the Water Quality Improvement
Act, first declared that states would be permitted to play a role in this
area. For by 1978, a Library of Congress Congressional Research Service
study showed that 18 coastal states had enacted their own oil spill laws.
A copy of the Library's study in chart form is reproduced below, showing the states which have enacted such legislation and the substantive
differences between the various enactments. Given the *Askew* decision,
the industry can look only to Congress and the courts for any real hope
of superseding these State enactments and achieving some degree of
uniformity on the basis of the doctrine of federal supremacy.

The Congressional debate over pre-emption is a lively one, reflecting
the range of views of the various interested outside parties. The Committee on Environment prefers no pre-emption, finding that even though
its reported Bill was 'the most stringent of the three "superfund" bills
in the Congress', were it to pre-empt State statutes it would impose 'a
lower standard of liability in one or more areas than virtually every State
statute'. Close examination of the argument for pre-emption, so the
Environment Committee concluded:

> reveals it as an argument rejected as flawed and dangerous in the
> adoption of the Federal system 200 years ago. Neither the circum-
> stances nor dangers have changed so much since that time that the
> Committee is now willing to embrace an authoritarian Federal regime.

The Senate Commerce Committee, expanding on comparable
provisions in the 1978 OCS Amendments[84] is probably inclined to
authorise pre-emption in at least two contexts: (1) precluding required
contributions to any State funds whose purpose is to pay compensation

Table 2: Comparable State Oilspill Liability Provisions

	Ala.	Alas.	Cal.	Conn.	Fla.	Ga.	La.	Me.	Ma.	Mass.	Miss.	NJ	NY	N.Car.	Oreg.	Tex.	Va.	Wash.
Funds																		
$ - a contributing fund	×	×			$		$	$	$			$	$	×	×	×	×	×
Liability																		
Strict (for hazardous type substances)		×				×						×		×	×		×	
Unlimited	×	×	×		×	×	×	×	×	×	×	×	×	×	×	×	×	×
Limited (except for negligence)				×	×							×				×		
Hazardous Substances																		
Includes hazardous substances		×			×	×						×				×		
Wide range of 'pollutants'	×		×	×							×				×			
Oil only							×	×	×	×				×		×	×	×
Damages																		
Clean-up, containment, damages	×	×	×	×	×	×	×	×	×	×				×	×	×	×	×
Restoration/ Restock	×		×								×		×		×			
Economic												×						

Source: S. Rep. No. 96-1152 (Committee on Environment and Public Works), 95th Cong., 2d Sess. 24 (1978).

for losses that can be compensated under the federal Act; and (2) precluding States from requiring any evidence of financial responsibility beyond that required under the federal Act. But, adopting language similar to that in the OCS Act, the Commerce Committee is still not prepared to pre-empt 'the field of liability or to preclude any State from imposing additional requirements or liability for damages and cleanup costs, within the jurisdiction of such State, resulting from a discharge of oil'. It cannot be said that this 'field' of non-pre-emption is a model of clarity.

For its part, the House prefers general pre-emption but goes about trying to achieve it not by outlawing or invalidating State laws on the subject, but by making the superfund and the federal courts the exclusive means for the collection of oil pollution damages.

9 CONCLUSION

Building upon the basis of the Shipowner's Limitation of Liability Act of 1851, US law regarding liability for oil pollution has proceeded at an irregular pace, providing what may fairly be characterised as a patchwork scheme of liability limits, legal defences and compensation fund programmes. The pattern is further complicated by the uncertain limits placed by the Constitution upon the power of the States to legislate in an area that has, in important respects, been entered by the Congress. Although the existence of diverse liability regimes may be explained as a matter of history, and to some extent justified as a recognition of separable factual settings of different areas of industrial activity, the proliferation of federal legislative programmes through a process of accretion has probably reached the point where the law begins to lose coherence, much as the addition of excessive levels of icing may topple an otherwise appealing four-layer cake. Congress now appears to be on the brink of redoing the uppermost layers of the legislative cake, although the precise ingredients will remain unclear until the task is done. The issue that will then have to be addressed is the need to reconcile domestic and international oil pollution compensation expectations and programmes in a way that will give due recognition not only to the 'common heritage of mankind' concept as applied to the global commons, but also to the proper interests of coastal States in preserving their environments, maintaining a regulatory and legal climate that will foster continued international trade in oil, and compensating those who have been injured as a result of that commerce. American law currently attempts this reconciliation, but the drive towards a comprehensive approach is such that the present legislative arrangements should probably be considered essentially transitory.[85]

NOTES

1. See Lettow, *The Control of Marine Pollution*, in Environmental Law Inst., Federal Environmental Law 598 (Dolgin and Guilbert eds. 1974).
2. 9 Stat. 635 (1851), 46 U.S.C. ss. 181-9 (1976).
3. The exceptions, all applying to oil pollution damage, are discussed in later sections of this chapter.
4. See 26 Geo. 3c. 86 (1786).
5. See generally Gilmore and Black, *The Law of Admiralty*, 818-19, 821 (2nd edn, 1975).
6. *Norwich Co.* v. *Wright*, 80 U.S. (13 Wall.) 104 (1871).
7. *Place* v. *Norwich & N.Y. Transp. Co.*, 118 U.S. 468 (1885); *Dyer* v. *National Steam Navigation Co.*, 118 U.S. 507 (1885); *Thommessen* v. *Whitwell*, 118 U.S. 520 (1885).
8. *In re Barracuda Tanker Corp.*, 281 F.Supp. 228, 230 (S.D.N.Y. 1968), *reversed on other grounds and remanded*, 409 F.2nd 1013 (2d Cir. 1969). See Sisson, 'Oil Pollution Law and the Limitation of Liability Act', (1978) 9 J.M.L.C. 285. The compensation paid for *Torrey Canyon* damage has been estimated to have ranged between $8m and $15m. No definitive figure has ever been published to the author's knowledge.
9. Gilmore and Black, *Law of Admiralty*, 877.
10. *The Titanic*, 233 U.S. 718, 732-33 (1914).
11. *Petition of Chadade Steamship Co., Owners of S.S. Yarmouth Castle*, 266 F.Supp. 517 (S.D. Fla. 1967).
12. Following the *Yarmouth Castle* disaster in November 1965, the Johnson Administration sponsored comprehensive maritime reform legislation, including a specific provision to repeal s. 183 in so far as it established limitations of liability for personal injury and death. S. 3251, 89th Cong., 2nd Sess. (1966). The comprehensive legislation was passed except for the repealer. In the 90th Congress, a repealer was again introduced, but it too was not enacted. H.R.17254, 90th Cong., 2nd Sess. (1968), S.3600, 90th Cong., 2nd Sess. (1968). In the context of oil pollution liability, no effort has ever been made to repeal s. 183. For its part, the Senate Commerce Committee considers that the limitations of s. 183 are applicable to oil pollution damage. See that Committee's recent report on Oil Pollution Liability. S. Rep. No. 95-427 (*Senate Committee on Commerce, Science and Transportation*), 94th Cong., 1st Sess. (1977) (hereafter referred to as *Sen. Comm. Rep.*). See also note 29 *infra*.
13. These included the 1899 Refuse Act, 33 U.S.C. s. 407 (1976); see Lettow, *Control of Marine Pollution*, at 602 and nn. 31-2; the Oil Pollution Act of 1924, as amended (43 Stat. 604, 33 U.S.C. s. 431 *et seq.* (1970)); the Federal Water Pollution Control Act of 1948, as amended (62 Stat. 1155, 33 U.S.C. s. 466 *et seq.* (1970)); and the Water Quality Improvement Act ('WQIA') of 1970 (Pub. L. No. 91-224, 84 Stat. 91, 33 U.S.C. s. 1161 *et seq.* (1970)). For a concise history of United States oil pollution legislation prior to 1972, see Davis, 'The Ports and Waterways Safety Act of 1972' (1975) 6 J.M.L.C. 249, and Milstein, 'Enforcing International Law: U.S. Agencies and the Regulation of Oil Pollution in American Waters', (1975) 6 J.M.L.C. 273.
14. Pub. L. No. 92-500, 86 Stat. 816. See also 1972 Code, Cong. & Adm. News 951, 3668.
15. 33 U.S.C. s. 1251(a)(1) (1976).

16. 33 U.S.C. ss. 1321(a)(1)-(2) (1976).

17. 33 U.S.C. ss. 1321(b)(5)-(6) (1976); Publ. L. No. 95-576, s. 7, 92 Stat. 2467 (1978).

18. 33 U.S.C. ss. 1321(c)(1), (d) (1976).

19. In 1970, when federal oil pollution limits of liability were first adopted in the WQIA, they were at that time set at *the lesser* of $100 per ton or $14 million. See WQIA, s. 11(f)(1), reprinted at 1970 U.S. Code, Cong. & Adm. News 101. No matter the tonnage of the tanker, therefore, its maximum exposure for oil pollution liability under the 1970 Act was $14 million. In the Clean Water Act of 1977 (Pub. L. No. 95-217, 91 Stat. 1566), these limits were increased to their present figures – which effectively amount to a limit of $150 per ton subject to no ceiling. As justification for removing the $14 million ceiling of the WQIA, the Senate Report on the 1977 Act stated that the ceiling

> served no useful purpose, inadvertently subsidizing large tankers and thus enhancing their competitive position over smaller vessels. The $14 million limit in existing law is totally inadequate to deal with an oil spill of any magnitude from the size of tanker that is expected to be plying the waters of the United States.

S. Rep. No. 95-370 (Senate Committee on Environment and Public Works), 95th Cong., 2nd Sess. (1977). See also 1977 U.S. Code, Cong. & Adm. News 4326, 4389. For views as to how the 'strict liability' standard made its appearance in United States and international oil pollution law, see Bergman, 'No Fault Liability for Oil Pollution Damage', (1973) 5 J.M.L.C. 1, and Gold, 'Pollution of the Sea and International Law', (1971) 3 J.M.L.C. 13.

20. 33 U.S.C. s. 1321(f)(1) (1976).

21. 33 U.S.C. s. 1321(p)(1) (1976).

22. 33 U.S.C. s. 1321(f)(1) (1976).

23. 33 U.S.C. s. 1321(a)(12) (1976).

24. 33 U.S.C. s. 1321(f)(1) (1976).

25. *Id.* and 33 U.S.C. s. 1321(n) (1976).

26. Except for the limits of liability, the 1972 Federal Water Pollution Control Act Amendments create an almost identical scheme of fines and liability for owners or operators of both on-shore and off-shore facilities from which oil or any hazardous substance is discharged. Their respective limits of liability, however, are set at $50 million, having been raised to these amounts by the Clean Water Act of 1977. When originally adopted in the 1970 WQIA, the limits for on-shore and off-shore facilities were set at $8 million. See 33 U.S.C. ss 1321(b)(5)-(6), (f)(2)-(3), (g).

27. A provision adopted in the 1977 Clean Water Act and presently codified at 33 U.S.C. s. 1321(f)(4) specifically includes as part of clean-up costs 'any costs or expenses' of the federal 'or any State government in the restoration or replacement of natural resources damaged or destroyed' as a result of an unlawful discharge. The terms of this provision could substantially expand potential damages under the Act, though, of course, not beyond the prescribed limits of liability. See 1977 U.S. Code, Cong. & Adm. News 4326, 4389.

28. 411 U.S. at 328-9.

29. Ibid. at 332. Professors Gilmore and Black have since characterised Justice Douglas's opinion in *Askew* as 'a new high-water in judicial ambiguity', adding that all *Askew* tells us 'is that there is a great deal of litigation in store on these complex issues'. Gilmore and Black, *Law of Admiralty*, at 833-4. Professors Healy and Sharpe concluded that liability established by State statutes 'will be subject to limitation in accordance with the [1851] Limited Liability Act'. Healy and

Sharpe, *Admiralty Cases and Materials*, 752 (1974). See also Swan, 'American Waterways: Florida Oil Pollution Legislation Makes It Over First Hurdle', (1975) 5 J.M.L.C. 77. More recently, the Supreme Court of the United States struck down that part of a Washington State law which attempted to govern the design, size and movement of oil tankers in Puget Sound and set certain higher standards than those prescribed under federal law. Without mentioning *Askew*, the Court held that the State law was pre-empted by federal law. *Ray* v. *Atlantic Richfield Co.*, U.S., 55 L.Ed.2d 179 (1978). For recent litigation in the lower federal courts on the pre-emption question, see Case Notes, (1979) 10 J.M.L.C. 287, (1978) 9 J.M.L.C. 397. In its 1978 Report on Oil Pollution Liability (Sen. Comm. Rep., *supra* note 12 at pp. 6 and 10), the Senate Commerce Committee referred to the 1970 and 1972 Water Pollution Acts but pointed out that '[n]othing has been done to improve the position of the injured property owner whose recovery of damages is still curtailed by the [1851] Limitation Act'. Re-emphasising this same point, the Committee later concluded that:

> In a time of increasing oil transportation by ships and in an era of a growing tanker technology, recovery of private oil spill damage is still governed by the Limitation of Liability Act – a statute which was enacted over 120 years ago and which still purports to deal with a problem which was not even in existence at the time of its passage.

The views of the Commerce Committee on this divisive issue thus seem clear.
30. Pub. L. No. 93-153, 87 Stat. 584, 43 U.S.C. ss. 1651-5 (1976).
31. 43 U.S.C. ss. 1653(c)(1), (7) (1976).
32. 43 U.S.C. s. 1653(a)(1) (1976).
33. 43 U.S.C. ss. 1653(a)(1), (c)(2) (1976).
34. Ibid.
35. 43 U.S.C. s. 1653(a)(2) (1976).
36. The concept of such a fund was by this time familiar to the Congress because of its earlier examination of the fund proposed to be created by the 1971 International Convention on the Establishment of an International Fund for Compensation for Oil Pollution Damage. In addition, at least ten United States coastal States had by this time adopted legislation establishing comparable oil pollution funds. See *infra*.
37. 43 U.S.C. ss. 1653(c)(3), (5) (1976).
38. S. Rep. No. 95-1152, 95th Cong., 2nd Sess. 15 (Committee on Environment and Public Works) (1978) (hereinafter referred to as *Sen. Envir. Rep.*).
39. 43 U.S.C. s. 1653(d)(9) (1976).
40. 43 U.S.C. s. 1653(c)(3) (1976).
41. Exec. G, 91st Cong., 2nd Sess. (1970).
42. Pub. L. No. 93-248, 88 Stat. 8. See also 1974 U.S. Code, Cong. & Adm. News 2773; Bissell, 'Intervention on the High Seas: An American Approach Employing Community Standards', (1976) 7 J.M.L.C. 718. It will remain an interesting, but unanswered, question in American law why all international maritime law conventions, whether of a private or public law type, are routinely implemented by a separate enactment into domestic law of the terms of the Convention, while international aviation law conventions (e.g. the Chicago Convention on International Civil Aviation, 61 Stat. 1180; the Warsaw Convention on the Liability of international air carriers, 49 Stat. 3000) are routinely deemed to be self-executing and thus to require only the most minimal domestic implementing legislation and often none at all.
43. 14 U.S.C. s. 3 (1976); 49 U.S.C. s. 1655(b)(2) (1976).
44. 33 U.S.C. ss. 1472-3 (1976).
45. 33 U.S.C. s. 1474(3) (1976). No similarly precise authorisation appears in

the language of the Convention, but Articles I, III and V are sufficiently broad as easily to encompass such acts as removing or, if necessary, destroying the vessel. The power to destroy vessels presenting hazards to navigation or risk of pollution was previously found in a variety of federal laws. See, for example, 14 U.S.C. s. 88(a) (1976); 33 U.S.C. s. 415 (1976).

46. 33 U.S.C. ss. 1477 & 1479 (1976).

47. 33 U.S.C. s. 1486 (1976). See note 23 *supra*.

48. Pub. L. No. 93-627, 88 Stat. 2126. See 1974 U.S. Code, Cong. & Adm. News 2468, 4529.

49. 33 U.S.C. ss. 1502 (16), 1509 & 1517 (1976). See Convention on the Continental Shelf, Art. 5, Apr. 29, 1978, T.I.A.S., No. 5578, 15 U.S.T. 471, 49 U.N.T.S. 311.

50. 33 U.S.C. ss. 1517(a)-(b) (1976).

51. 33 U S.C. ss. 1517(d)(e) & (g) (1976).

52. 33 U.S.C. s. 1517(d) (1976).

53. 33 U.S.C. s. 1517(e) (1976).

54. 33 U.S.C. ss. 1517(d)-(e) (1976).

55. 33 U.S.C. s. 1517(f) (1976).

56. 33 U.S.C. ss. 1517(i)(1), (3) (1976).

57. For helpful reviews of the background of these events, see Wood, 'An Integrated International and Domestic Approach to Civil Liability For Vessel-Source Oil Pollution', (1975) 7 J.M.L.C. 1 and Goldie, 'Liability For Oil Pollution Disasters', (1975) 6 J.M.L.C. 323.

58. See *infra*.

59. 67 Stat. 462, 43 U.S.C. ss. 1333 *et seq.* (1976).

60. Pub. L. No. 95-372, 92 Stat. 629. See also 1978 U.S. Code, Cong. & Adm. News 629, 2856.

61. H.R. Rep. No. 95-590 (OCS Committee), 95th Cong., 2nd Sess. 65 (1978). See also 1978 U.S. Code, Cong. & Admn. News 2878.

62. Ss. 301, 304(a), 304(b)(1)-(2); 43 U.S.C. ss. 1811 & 1814.

63. 33 U.S.C. ss. 1321(c), (d) & (1).

64. 33 U.S.C. s. 1474.

65. Ss. 301(22), 303(a)(1) & 304(d); 43 U.S.C. s. 1811-13.

66. S. 304(b), 43 U.S.C. s. 1814 (emphasis added).

67. S. 304(c)(1), 43 U.S.C. s. 1814.

68. 33 U.S.C. s. 1321(f)(i).

69. Compare notes 33, 34 and 51 *supra*, and accompanying text.

70. Ss. 303(a)(1) and (a)(2). For the bracketed qualifications, see ss. 303(b)(4) and (b)(5); 43 U.S.C. s. 1813. The powers conferred on the President with respect to the assertion of oil pollution claims under the 1978 Amendments have been delegated to the Secretaries of Commerce and the Interior. Exec. Order No. 12123, s. 1-1, 44 Fed. Reg. 11199 (28 February 1979). Other powers relating to the determination of financial responsibility and the assessment and compromise of penalties were delegated to the Federal Maritime Commission and the Secretary of Transportation. Ibid., s. 1-2.

71. S. 304(g); 43 U.S.C. s. 1814 (1976).

72. Ss. 302(a), (d); 43 U.S.C. s. 1812 (1976).

73. Ss. 302(a), (f), 43 U.S.C. s. 1812 (1976).

74. S. 304(f)(1), 43 U.S.C. s. 1814 (1976). Section 304(f)(2) exonerates the fund from any liability to a claimant whose negligence or wilful misconduct caused the damage for which the claim is asserted.

75. S. 305, 43 U.S.C. s. 1815. Such regulations were published by the Federal Maritime Commission and the Coast Guard in March 1979.

76. S. 312, 43 U.S.C. s. 1822.

77. US Dept. of Justice, *Methods and Procedures for Implementing a Uniform*

Law Providing Liability for Clean up Costs and Damages Caused by Oil Spills from Ocean Related Sources, 94th Cong., 1st Sess. (Comm. Print, 1975).

78. See materials cited at note 83 *infra*.

79. For the sources of these comparisons, see *Sen. Comm. Rep., supra* note 12; *Sen. Envir. Rep., supra* note 38; and H.R. Rep. No. 95-340 (House Committee on Merchant Marine and Fisheries), 95th Cong., 1st Sess. (1977).

80. 33 U.S.C. s. 1321(f)(1) (1976).

81. 33 U.S.C. ss. 1321(f)(2)-(3) (1976).

82. The $39 and $78 million amounts reflect the increase in gold value since adoption of the Fund Convention. It is also virtually certain that the United States will not ratify the CLC, considering its limit of $134 per gross ton (subject to a $14 million ceiling) compared with the limit of $300 per ton (with no ceiling) which has been adopted in the 1978 OCS Act and which both responsible Senate Committees seem prepared to accept today in any superfund legislation. Moreover, in its 1978 Report, the Senate Commerce Committee expressly concluded (at p. 4) that: 'Because of the inadequacies of [the 1969 and 1971] proposed treaties the Committee believes they should not be approved unless substantially altered. In the meantime, domestic legislation should fill the gap.' This view, or one very much like it, is probably shared by a majority of the present members of the Senate Committee on Environment. As a result, unless there is a radical change in the way the international community views the matter of compensation for oil pollution damage, it is hard to envision a successful leadership role for the United States in future treaty-writing efforts to unify the international law governing oil pollution liability. The pending Third United Nations Conference on the Law of the Sea, if it leads to conclusion of a new Convention, may spark renewed international activities with respect to pollution of the sea by oil. Because the outcome of the Conference is still a matter of speculation, however, the authors have thought it best simply to note the possibility that changes in domestic legislation in the United States could be among the results of such a new international accord.

83. See 125 Cong. Rec. s. 1013-16 (daily ed. 1 February 1979); 124 Cong. Rec. (daily ed. 14 October 1978).

84. S. 310, 43 U.S.C. s. 1820 (1976).

85. Cf. Lettow, *Control of Marine Pollution*, at 629.

NOTES ON CONTRIBUTORS

Gordon L. Becker, Counsel, Law Department, Exxon Corporation, 1251 Avenue of the Americas, New York, N.Y. 10020, U.S.A.

D. Alastair Bigham, Barrister, F.R.I.C.S., F.I. Arb., 2 Harcourt Buildings, Temple, London EC4Y 9DB.

Mrs Patricia Birnie, B.A., Barrister, Lecturer in Public International Law, University of Edinburgh, Old College, South Bridge, Edinburgh EH8.

Robin Churchill, LL.M., Lecturer in Law, U.W.I.S.T., King Edward VII Avenue, Cardiff CF1 3NU.

Douglas J. Cusine, LL.B., Solicitor, Lecturer in Private Law, University of Aberdeen, Taylor Building, Old Aberdeen AB9 2UB.

Eugene R. Fidell, Partner in the Law Firm of LeBouef, Lamb, Leiby & MacRae, 1333 New Hampshire Avenue N.W., Washington D.C. 20036, U.S.A. Member of the Editorial Board of the *Journal of Maritime Law and Commerce.*

Carl August Fleischer, Professor of Law at the University of Oslo, Special Consultant to the Norwegian Ministry of Foreign Affairs in Matters of International Law, Head of the Norwegian Delegation and Vice-President at the 1975 and 1976 London Conference on Civil Liability for Oil Pollution Damage resulting from Offshore Operations, etc.

Angelantonio D.M. Forte, LL.B., Solicitor, Lecturer in Private Law, University of Glasgow, Glasgow G12 8QQ.

John P. Grant, LL.B., LL.M., Senior Lecturer in Public International Law, University of Glasgow, Glasgow G12 8QQ.

Dr Philip Kunig, Institut für Internationale Angelegenheiten, Universität Hamburg, Rottenbaumchaussee 19-23, 2000 Hamburg 13.

S. Mankabady, LL.M., Ph.D., Reader in Shipping Law, Liverpool Polytechnic, 8 Hall Road East, Blundellsands, Liverpool L23 8ST.

Allan I. Mendelsohn, Partner in the Law Firm of Ward and Mendelsohn,

1715 Eye Street N.W., Washington D.C. 20006, U.S.A. Member of the Editorial Board of the *Journal of Maritime Law and Commerce.*

INDEX

Abatement notice 212
Abuse of rights 31-2, 183-4
Accidents (to tankers)
  attempts to eliminate *see*
  Collisions, Tankers – Construction
  and equipment, Crew standards;
  efficacy of measures 39; scale of
  pollution 29, 35-6
Adair, Paul 'Red' 140-1
Advisory Committee on Oil Pollution
  of the Sea (ACOPS)
  and unidentified polluters 57, 60;
  criticism of government 61
Amoco Cadiz
  and IMCO 92; consequences 22-3;
  the ship 21; the stranding 21-2
Alkali and Clean Air Inspectorate (UK)
  211-12
Alkali etc. Act 1863 (UK) 211
Alkali etc. Works Regulation Act (UK)
  (1906) 211
Alphacell Ltd v Woodward 216
American Bar Association
  Annual Report 1974 219-20
Artic Waters Pollution Prevention
  Act 1970 (Canada) 30, 97, 183,
  243-4
*Askew v American Waterways
  Operations Inc.* 158, 297

Barcelona Convention for the Protection of the Mediterranean Sea
  against Pollution 1976
  comparison with Paris Convention
  for the Prevention of Marine
  Pollution from land-based sources
  1974 270-2; contracting parties
  269; enforcement 272; obligations
  271-2; pollution 270
Bathing Water
  European Community directive
  276-8
Blow-out
  blow-out preventer 46; insurance
  against 168-9; risk of 135-6
Brussels Convention on the Liability of
  Operators of Nuclear Ships 1962
  245

Canada
  Arctic Waters Pollution Prevention
  Act 1970 30, 97, 183, 243-4
Civil Liability
  Civil Liability Convention 1969:
  actual fault or privity 54; amendments 81-2; compared with
  TOVALOP 125-6; compromise
  in London Convention 149-50;
  contributory negligence 52;
  defences 52; enforcement of judgments 55; genesis of 50-1; insurance
  52, 55; jurisdiction under 51;
  liability 51-2, 156; limitation of
  liability 53-4, 156-7; oil, definition
  of 34; role of IMCO 81; substances
  other than oil 82; *Torrey Canyon*
  and 81
  In Norwegian Law 137
  In United Kingdom: actual fault
  or privity 54; contributory
  negligence 53; defects in common
  law 50-1; defences 52-3; enforcement of judgments 55; *Esso Petroleum Co. Ltd v Southport Corporation* 32-3, 51; insurance 55-6;
  jurisdiction 51; liability 52-3;
  liability fund 54; limitation on
  liability 54; Merchant Shipping
  (Oil Pollution Act) 1971 51; Scots
  Law 33; *Torrey Canyon* 50-1
Clean Water Act 1977 (USA)
  Administrator of the Environmental Protection Agency 223-35;
  area wide treatment and management schemes 224-5; *Askew v
  American Waterways Operators Inc.*
  297; clean-up 295-6; Committee
  of Public Works and Transportation
  226, 235; effluent limitations
  226-7; Effluent Standards and
  Water Quality Information
  Advisory Committee 234-5, 236;
  Federal enforcement 231-4; Federal
  Water Pollution Control Act 1972
  222-35, 295-7; guidelines and
  criteria 229-30; liability of federal
  government 233-4; liability under
  295-6; management practices 225;
  National Contingency Plan 232;
  National Pollutant Discharge
  Elimination Scheme 234; national

318